Bioenergy and Biological Invasions: Ecological, Agronomic and Policy Perspectives on Minimizing Risk

Bioenergy and Biological Invasions: Ecological, Agronomic and Policy Perspectives on Minimizing Risk

Editor

Ajay Deshpande

Bioenergy and Biological Invasions: Ecological, Agronomic and Policy Perspectives on Minimizing Risk

Edited by **Ajay Deshpande**

Printed in 2017

ISBN: 978-1-68117-048-0

Library of Congress Control Number: 2015935443

© 2016 by
SCITUS Academics LLC,
616, Corporate Way, Suite 2, 4766,
Valley Cottage, NY 10989

www.scitusacademics.com

Contents

Robles, J. Lakatos, P. Scharek, Z. Planck, G. Hernández, S. Solís,
and E. Bustos

J. A. García, D. Monzón, A. Martínez, S. Pamukcu, R. García,
and E. Bustos

Preface

The emerging bioeconomy offers numerous potential benefits, including development of carbon-neutral energy sources, energy independence, production of novel bioproducts, and renewal of rural economies. Aside from the benefits, the emerging bioeconomy is likely to result in the single largest reconfiguration of the agricultural landscape since the advent of industrial agriculture. The scale and pace of this revolution pose significant challenges for sustainable bioeconomic development. We present and explore some of the key ecological and environmental challenges of one aspect of the bioeconomy - biofuel production. In assessing these challenges, we highlight the inadequacy of seeking simplistic solutions. We identify the need for a multidimensional approach to addressing these challenges. We propose that the framework of 'biocomplexity' enables such a multidimensional and cross-disciplinary consideration of biofuel production. Integration of such a systems approach to biofuel development, with a more-inclusive public engagement process, would be beneficial within a triple bottom line context. Despite major international investment in biofuels, the invasive risks associated with these crops are still unknown. A cohesive state-of-the-art review of the invasive potential of bioenergy crops, this book covers the identified risks of invasion, distributions of key crops and policy and management issues. Including a section on developing predictive models, this book also assesses the potential societal impact of bioenergy crops and how to mitigate invasive risks.

Editor

Plants as Useful Vectors to Reduce Environmental Toxic Arsenic Content

Nosheen Mirza[1], Qaisar Mahmood1, Mohammad Maroof Shah[1], Arshid Pervez[1], and Sikander Sultan[2]

[1]Department of Environmental Sciences, COMSATS Institute of Information Technology, Abbottabad 22060, Pakistan

[2]Department of Microbiology and Molecular Genetics, University of the Punjab, Quaid-i-Azam Campus, Lahore 54590, Pakistan

ABSTRACT

Arsenic (As) toxicity in soil and water is an increasing menace around the globe. Its concentration both in soil and environment is due to natural and anthropogenic activities. Rising arsenic concentrations in groundwater is alarming due to the health risks to plants, animals, and human beings. Anthropogenic As contamination of soil may result from mining, milling, and smelting of copper, lead, zinc sulfide ores, hide tanning waste, dyes, chemical weapons, electroplating, gas exhaust, application of municipal sludge on land, combustion of fossil fuels, As additives to livestock feed, coal fly ash, and use of arsenical pesticides in agricultural sector. Phytoremediation can be viewed as biological, solar-driven, pump-and-treat system with an extensive, self-extending uptake network (the root system) that enhances the natural ecosystems for subsequent productive use. The present review presents recent scientific developments regarding phytoremediation of arsenic contaminated environments and its possible detoxification mechanisms in plants.

INTRODUCTION

Arsenic is a trace metalloid found in almost all environments. It exists in the −3, 0, +3, and +5 oxidation states. Environmental forms include arsenious acids, arsenic acids, arsenites, arsenates, methylarsenic acid (MAA), dimethylarsinic acid (DMAA), trimethyl arsine oxide (TMAO), and so forth [1–3]. The most abundant forms of arsenic include arsenate (As V) and arsenite (As III) [4], where arsenite is a toxic and hard acid. Arsenate (As V) usually forms complexes with sulfides, whereas arsenite (As III) develops complexes with oxides and nitrogen chemical species. Under both the oxidized and reduced states, As is sensitive to mobilization at pH range of 6.5~8.5 [5].

Arsenic toxicity of soil and water is an increasing menace across the globe [5, 6]. Millions of people especially in developing countries of Southeast Asia and many other regions are chronically exposed to As contamination [7, 8]. As carcinogenicity is well documented as it seriously affects human health and causes bladder, lung, and skin cancers and possibly damage to liver and kidney as well [7, 8]. Noncancerous health effects of As exposure include diabetes, skin diseases, chronic cough, and toxic effects on liver, kidney, cardiovascular, and nervous system [7–11]. Moreover, As contamination has become a major environmental concern because it not only adversely affects humans but also causes highly toxic effects on metabolic processes of plants, mitotic abnormalities, leaf chlorosis, growth inhibition, reduced photosynthesis, DNA replication, and inhibition or activating enzymatic activities [12].

ARSENIC CONCENTRATIONS IN SOIL AND WATER

Arsenic is the 20th most abundant element in earth crust [13], making about $5\,mg\,kg^{-1}$ of earth's crust, with an average concentration of 2 $mgkg^{-1}$ in igneous and sedimentary rocks [14]. It is a naturally occurring element typically found in soil at background concentrations ranging from 0.1 to 40 $mgkg^{-1}$. As is commonly associated with sulfides, oxides/hydroxides of aluminum (Al), iron (Fe), and manganese (Mn); other sources are volcanic eruptions and sea salt sprays [15]. In soil, As is present in the form of oxides, hydroxide, chlorides, and sulfides, such as enargite (Cu_3AsA_4), cobaltite (CoAsS), and skutterudite ($CoAsS_4$) and its average concentration in different regions of the world is 9.36 $mgkg^{-1}$. Heavy use of As containing pesticides is considered as the major reason for its pollution [16]. Arsenic and P are chemically similar. Both form insoluble compounds with Al and Fe in soils. In soil, Al-As and Fe-As complexes are the dominant chemical forms, while arsenic has less affinity for Al oxides than phosphates. As (III) gets adsorbed on iron (III) surfaces [17]. Kaolinite and montmorillonite have higher affinities for As (V) than for As (III) [18]. Arsenic mobility and phytotoxicity are greater in sandy soils.

Rising arsenic concentrations in groundwater are alarming due to the health risk to plants, animals, and humans health [19]. Higher levels of arsenic were found in groundwater sources than in surface-water sources. Many countries around the world (including Taiwan, Argentina, India, Bangladesh, Mexico, Hungary, and Chile) have reported extensive arsenic groundwater contamination [19, 20]. Use of such contaminated water for irrigation of crops may lead to arsenic contamination of agricultural soils. The presence of high As concentration in the aquifer may be due to desorption of arsenic from Fe and Mn oxides, weathering of primary silicate minerals, and apatite under high pH and alkalinity from silicate and carbonate reactions [21].

Anthropogenic Sources of Arsenic

Anthropogenic As contamination of soil may result from mining, milling, and smelting of copper, lead, and zinc sulfide ores, hide tanning waste, dyes, chemical weapons, electroplating, gas exhaust, municipal sludge of land, combustion of fossil fuels, As additives to livestock feed, coal fly ash, and agricultural use of arsenical pesticides [3, 22–30]. In the past decade, the global input of As to soils by human activities was estimated to be around 52,000–112,000 ton per year [31]. Thus, the arsenic concentrations in soil and environment both are due to natural and anthropogenic activities. Most of the arsenic risk is associated with the forms that are biologically available for absorption or "bioavailable" to plants and humans. A bioavailable chemical is the portion of a chemical dose that enters the systemic circulation from an administered dose [32].

Of 1.4 million worldwide contaminated sites 41% are in the USA and US EPA has recognized that arsenic (As) concentration in Australia was greater than 10,000 mgkg^{-1} [33]. Arsenic has been found at high levels (10000–20000 mgkg^{-1}) in some contaminated areas and that results in unacceptable levels of risk to human health from the incidental ingestion of soil [34]. Groundwater arsenic contamination has been reported in many parts of the world, such as Vietnam, Massachusetts State, Carolina State, Canada, and Bangladesh, with 0.305, 30, 2460, 6590, and 0.3990 mgkg^{-1} arsenic (As) contamination [35–40]. As intake through drinking water is a very severe problem in the Southeast Asia with the Bengal delta being the worst affected area [41]. Mining has resulted in increased As concentrations in Warsak Canal [42]. Large arsenic concentrations such as 0.942, 0.40, 0.38, 0.643, and 0.475 mgL^{-1} were found in Hattar Industrial Estates, Ghari Rahimabad, Pakha Ghulam, Peshawar Industrial Estate, and Gujranwala Industrial Estate, respectively, in Pakistan [43].

One of the more widespread problems is the leaching of naturally occurring arsenic into drinking water aquifers [3]. Thus, groundwater As contamination is the most common result of its higher concentrations in soil. It is estimated that approximately one third of the world's population use groundwater for drinking [44, 45], which ultimately adversely affects human beings as the biggest calamity, was in Bangladesh, where millions of people were dependent on

As contaminated drinking water [12] and it is the possible cause of the death of such notables as Napoleon and the American president Zachary Taylor [3].

The reduction of the World Health Organization (WHO) provisional guideline value for As concentration in drinking water was from 50 μgL^{-1} to a provisional 10 μgL^{-1} in 1993 [46]. However, only during the past 5 years many industrial countries adopted that lowered guideline value as the maximum contaminant level (MCL). On the other hand many developing countries including India and Bangladesh still have 50 μgL^{-1} as MCL [41] and the reduction of the maximum admissible concentration (MAC) to 10 μgL^{-1} by USEPA in 2002 was a response to growing concern over that poisonous carcinogen which raised awareness of the dangers of As in drinking water [47]. In view of the health concerns outlined above, and alerted by the magnitude of the problem afflicting nearby Bangladesh and West Bengal, the Public Health Engineering Department (PHED), the Local Government and Rural Development Department (LGRDD) of Pakistan, in conjunction with UNICEF, undertook a survey of As concentration in groundwater from drinking water supply wells in Pakistan [48]. That survey revealed hot spots of As enrichment in parts of the Indus alluvial basin. The survey identified Muzaffargarh District (Pakistan) as one enriched in As at concentrations in the low hundreds of μgL^{-1} range. During the investigation, the authors found As "cold spots;" that is, areas where evaporative concentration of groundwater might have been expected to result in high concentrations of As in groundwater, but where concentrations were, in fact, below levels of concern [47].

Soil contamination and groundwater can be due to industrial point sources, repeated use of metal enriched fertilizers, farm manuring, sewage sludge, pesticides application, mining, automotive emissions, dyestuffs, and wood preservation [49]. As arsenic concentrations above acceptable standards have been detected in many countries such as Bangladesh, Cambodia, China, Taiwan, Inner Mangolia, India, Iran, Japan, Nepal, Pakistan, Thailand, Vietnam, Alaska, Argentina, Chile, Mexico, United States of America, Austria, Finland, France, Germany, Greece, Italy, Russia, United Kingdom, South Africa, Australia, and New Zealand [50]. In some areas of the Pakistan, the presence of arsenic in subsurface aquifers and drinking water systems is a potentially serious human health hazard. A majority of shallow subsurface aquifers and tube wells are contaminated with arsenic at levels which are above the recommended arsenic level of 10 ppb.

Soil and water contamination can be removed by immobilization, vitrification, soil washing/flushing, precipitation, membrane filtration, adsorption, ion exchange, permeable reactive barriers biological treatment, thermal processes, excavation and disposal process, chemical processes, and phytoremediation costing 75–425, 100–500, 100–500, and 5–40 dollars per ton of soil, respectively [51]. Most of these methods are found very costly, whereas phytoremediation has been suggested as the most cost effective and efficient method for removal or minimization of metal contamination both in soil and water [52]. Phytoremediation was firstly proposed over 20 years ago and is advantageous over chemical stabilization, which may prevent health threats occurring due to leakage of toxic metals [53]. Phytoremediation has also been called green remediation, botanoremediation, agroremediation, and vegetative remediation. It is a natural process of growing plants to remediate soil and water without affecting the landscape. Phytoremediation utilizes biological processes and anatomy and physiology of plants. It is plant-based soil remediation system can be viewed as biological, solar-driven, pump-and-treat systems with an extensive, self-extending uptake network (the root system) that enhances the below-ground ecosystem for subsequent productive use [54, 55]. Phytoremediation is a continuum of processes occurring to varying degrees under different conditions, media, contaminants, and plants [54]. Plants have both constitutive (present in most phenotypes) and adaptive mechanisms (present only in tolerant types) to cope with the elevated metal concentrations [51, 56]. They can absorb and accumulate metals much higher than they need. The metals are generally accumulated in their aerial tissues [57].

ROLE OF PHYTOREMEDIATION

Numerous terms are being used simultaneously in the literature to refer to these processes and may overlap to some extent. Phytoremediation consists of four to five different technologies [54, 55], each having a different mechanism such as the following.

(1)   Phytoextraction or phytomining or phytoaccumulation: plants take up and translocate metal contaminants from soil to the

above ground portions, which then are harvested to remove the contaminant from the site.

(2)   Phytodegradation or phytotransformation: plants disintegrate pollutants which may occur within the plant by the metabolic activity or breakdown of the pollutant external to the plant contributed by various organic compounds released into the rhizosphere.

(3)   Rhizofiltration: plants get rid of contaminants present in solution surrounding the root zone by adsorption or precipitation onto their roots or absorption of contaminants into their roots from the solution. This technique is used to clean contaminated water such as groundwater or a waste stream.

(4)   Phytostabilization: plants immobilize contaminants in the soil and groundwater through absorption and accumulation by root or precipitation within the rhizosphere.

(5)   Phytovolatilization: plants volatilize pollutants; they take up the pollutants from the soil or water in the transpiration stream and volatilize into the atmosphere in a modified or unmodified form.

Arsenic phytoremediation involves immobilization, fixation, and removal either as fixed in soil or accumulated in plant parts.

Role of Plants in Remediation of Arsenic

Plants require an adequate supply of all nutrients, as part of normal growth and development [58], including arsenic, for their normal physiological and biological functions. Deficiency of specific nutrient occurs when plants cannot obtain sufficient amount as required, whereas excessive supply of the same, through contaminated soil results in toxicity to plants. Recommended soil application by US EPA for arsenic (As) is 41 $mgkg^{-1}$, whereas recommended standards by WHO for drinking water and effluents to be released by industries are 1, 0.01, and < 0.01 mgL^{-1}. The global input of arsenic to soils by humans in the last decade was estimated between 52,000 and 112,000 t $year^{-1}$ [31]. Arsenic contaminated sites can be remediated by utilizing the ex situ physical and chemical techniques [51]. But physicochemical remedies render the land futile for further use, during the process of decontamination, since they abolish all biological activities contributed by beneficial microorganisms, which are necessary for plant growth and development. Consequently, the ecosystems deteriorate with a decline in biodiversity. Arsenic contaminated sites usually have adverse soil conditions, that is, poor soil structure, low organic content, inadequate N and P, and so forth, and plants need to adapt to these hostile soil conditions as well as to the metal contamination.

Generally, prior to imposed selection, a species must be able to thrive and survive in As contaminated soil and/or water, for which it must possess appropriate variances [59]. Thus, only plants possessing tolerance show some preadaptation to these harsh conditions. Notable examples of such plants are Andropogon scoparius, ribwort plantain (Plantago lanceolata L., Holcus lanatus),mosses, lichens, crowberry (Empetrum nigrum L.), Tamarix (Tamarix parviflora), Eucalyptus (Eucalyptus camaldulensis), and Chinese Brake fern (Pteris vittata L.) [59–68]. Tolerance of plants to metals is under control of uptake systems which are directly related to metal concentrations in the soil solutions. Plants mostly possess two uptake systems: the highly inducible high-affinity system operational at low concentrations (such as the high affinity phosphate uptake system under low phosphate status) and the constitutive low-affinity system that is effective at high concentrations [51, 61, 69–71]. For uptake, arsenic needs to be bioavailable. Two mechanisms are responsible for arsenic transport from the bulk soil to plant roots, mass flow, and diffusion. Thereafter, plants may utilize two separate systems to take up arsenic: (1) passive uptake through the apoplast and (2) active uptake through the symplast [51]. Once arsenic is taken up, it is translocated from the roots to the shoot system via the xylem and redistributed between tissues. The translocation of arsenic and other metals depends upon root pressure and leaf transpiration [51, 72]. Most plants take up arsenic as arsenate [73] since arsenite is unstable as it gets oxidized to arsenate by biochemical processes in the soil system. Arsenate being a chemical analogue of phosphate competes with phosphate for its uptake system and is actively taken up [61, 74, 75]. Once taken up, it is reduced in the cytosol to arsenite by glutathione (GSH) [76] and translocated to the shoots [77, 78].

Generally, only a minuscule amount of arsenic is translocated to the aboveground parts leading to little accumulation. The form in which arsenic translocated in plants was unknown until 1999 [79]. There was some evidence that arsenic transported as dimethylarsenic acid to the shoots [80] and may be stored as an arsenite-tris-thiolate complex [81] in tissues [82].

Detoxification Mechanisms in Plants

Large green plants have the capability to move large amounts of soil solution into the plant body through the roots and evaporate this water out of the leaves as pure water vapour during transpiration. Plants transpire water to move nutrients from the soil solution to leaves and stems, where photosynthesis occurs, and to cool the plant. During this process, contaminants present in the soil water are also taken up and sequestered, metabolized, or vaporized out of the leaves along with the transpired water.

Heavy metals are generally transported and deposited in the vacuole as metal chelates. According to Baker et al. [83], free metal ions in the solution are taken up by plants into their tissues and are reduced as metal chelates using specific high-affinity ligands (like oxygen-donor ligands, sulfur-donor ligands, and nitrogen-donor ligands), for example, carboxylic acid anions which are abundant in the cells of terrestrial plants and form complexes with divalent and trivalent metal ions of reasonably high stability. Carboxylates (such as malate, aconitate, malonate, oxalate, tartrate, citrate, and isocitrate) are commonly the major charge-balancing anion present in the cell vacuoles of photosynthetic tissues and several of these carboxylates get associated with high metal concentrations in plants [84–86].

Sulfur-donor ligands (like metallothioneins and phytochelatins) form highly stable complexes with heavy metals because sulfur is a better electron donor than oxygen. Metallothioneins are gene-encoded low-molecular-weight, cysteine-rich peptides found in fungi and mammals recently shown to be induced by Cu [87]. In fungi and mammals, metallothioneins are involved in metal detoxification [88] but their role in plants is not yet well understood.

Plants employ several extracellular and intracellular mechanisms to detoxify heavy metals [51, 89]. These mechanisms include chelation, compartmentalization, biotransformation, and cellular repair [90]. The external mechanisms include exudations which change rhizosphere pH, metal speciation, and binds metal ions on the cell walls. Intracellular mechanisms include alteration of cell membrane or other structural protein to reduce the effects of metal toxicity and ultimate transport of metal to vacuole where detoxification occurs. Detoxification at the cellular level involves subcellular compartmentalization, chelation of metal in the cytosol by high affinity ligands, or binding metals to cell walls. Figure 1 shows the possible As accumulation and volatilization in Arundo donax L.

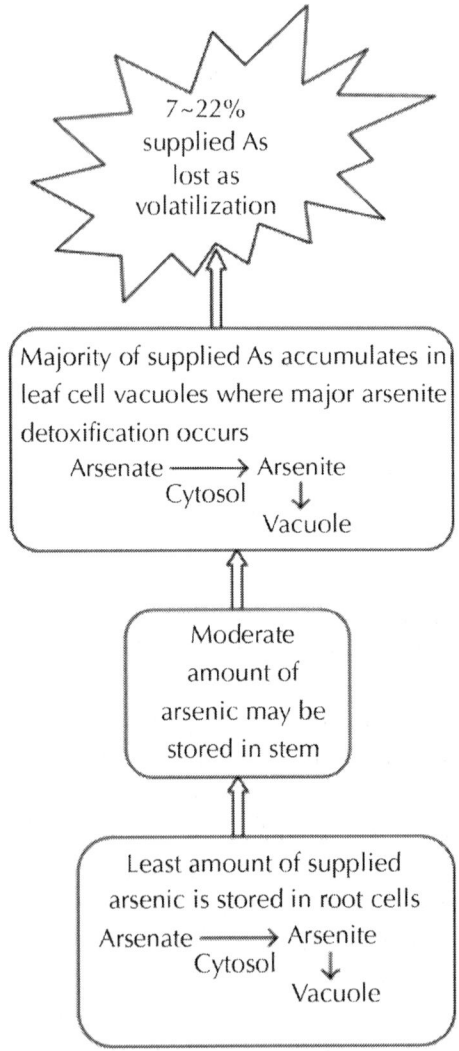

Figure 1: The overview of arsenic accumulation during phytoremediation experiments as suggested by the recent studies of Mirza et al. [91] and Doucleff and Terry [92]. It was suggested that arsenic is absorbed as arsenate and some part of it is converted into arsenite by an enzyme called arsenate reductase.

Arsenite is stored in vacuole and is further detoxified. Majority of arsenate is transported to leaf cells where it is again converted into arsenite (in cytosol) and stored/detoxified in vacuoles. Still majority of investigators have proposed that arsenite form is transported from roots to shoots.

For many contaminants, passive uptake via micropores in the root cell walls may be a major route into the root, where sequestration or degradation occurs. The apoplast is a hydrated free space continuum between the external soil solution and the cell membranes of the root cortex and vascular tissue. The cell wall micropores exist within a network of cellulose, hemicelluloses, pectins, and glycoprotein containing many negative charges (generated by carboxylic groups) that act as cation binding sites and exchangers and as anion repellers. Di- and polyvalent cations (the form of many heavy metal and radionuclide contaminants) are preferentially attracted to, and bound on, these cation exchange sites within the root cortex cell walls. For metal ions to be metabolized or translocated to the aboveground parts of the plant, they must pass through the plasma membrane of a living cell, and this can only occur by active transport processes. The inner limit of the root cell wall is the endodermis, which forms the outer limit of the root vascular system or stele.

Phytochelatins are low-molecular-weight, cysteine-rich peptides that are especially produced by plants when exposed to heavy metals and are known to bind metal in plants [93]. The PC-metal complexes are less toxic than free metal ions to cellular plant metabolism. Phytochelatin synthesis has been induced on exposure to arsenate in a number of plant species [79, 94, 95]. Intact PCs-As complexes have also been isolated from plant tissues [94] suggesting that phytochelatins are also involved in arsenic detoxification in plants. Though phytochelatin (PC) synthesis was induced on exposure to arsenate in P. vittata, only PC2 was detected in the plant. The molar ratio of PC-SH to As suggested that only a small proportion (1–3%) of the As in P. vittata can be complexed with PCs [95].

The metal-binding peptides, such as thiol-rich phytochelatins, have been most widely found and studied in plants particularly in response to arsenic [96]. They provide a detoxification mechanism by arsenic to their thiol group [79, 82, 94]. Where phytochelatins are derived from glutathione (GSH) [97] their biosynthesis is from GSH due to the presence of phytochelatin synthase enzyme [98]. Arsenic detoxification by phytochelatins on exposure to arsenate firstly was suggested by Grill et al. [99]. As it is understood that the immobilized metals are less toxic than the free ions, thus binding of arsenic to phytochelatins is considered to be a part of the detoxifying mechanisms of higher plants. After complexion; that is, the phytochelatin-metal complexes, for example, Cd-phytochelatin complexes, are transported to the vacuole which may be the final storage compartment where they either dissociate, degrade, or are shuttled back into the cytoplasm due to the acidic vacuolar pH [82, 100, 101]. On the contrary, if arsenic-phytochelatin complexes are transported inside vacuole, they might remain stable and prevent re-oxidation of arsenite due to the acidic pH of the vacuole, allowing accumulation of high concentrations of arsenic phytochelatin complexes [82].

Chintakovid et al. [102] pointed out that plants can tolerate the toxicity of arsenic by inhibiting translocation to the shoots, thus accumulating it primarily in the roots. Chintakovid et al. [102] found higher arsenic concentration in roots than in shoots of cotton exposed for a short time to arsenic. Although the mechanism of arsenic accumulation in the stem is still unclear, the results suggested that arsenic was transported to the stems and the leaves of the nugget marigold via the vascular system. The samples showed high percentages of arsenite in stems and leaves while a high percentage of arsenate was found in the roots. Similarly, the relative distribution of As in plants shows that Brassica sp. accumulated As mainly in the roots followed by shoots and flower [102].

It was found that arsenite was the main arsenic species in the fronds. Both species of arsenite and arsenate were found in xylem sap from stems of Brassica juncea [81] and sunflower [103]. However, it was not known whether both species were actually loaded in the xylem sap or occurred as a result of the reduction and oxidation of As species during translocation in the xylem sap. Raab et al. [103] found arsenic-phytochelatin complexes in the roots, stems, and leaves of an arsenic nontolerant plant (Helianthus annuus) during the exposure to arsenite or arsenate. But in most cases, most of the arsenic (75–95%) in the fronds is present in the form of arsenite (3+ oxidation state) [68, 104, 105]. Arsenite was also the predominant form of arsenic in excised aerial tissues that was exposed to arsenic, whereas As (V) was the main form in excised roots suggesting that As (V) reduction occurred mostly in the fronds, mainly in the pinnae.

Detoxification Mechanisms in *Arundo donax* L.

A. donax is an erect, perennial, bamboo-like grass which has been present in the Mediterranean basin for thousands of years [106–108]. A. donax has become globally dispersed by humans, so it is possible to find it in Asia, south Europe, North Africa, the Middle East, and also in North and South America and Australasia [106–108]. A. donax can be used for many purposes in the rural world, such as lattices, fences, baskets, fishing rods, and stalks for plants, roofs, windbreaks, sun shelters, cereal bins, musical instruments, walking sticks, and trellises. It is the most widespread among the species of the genus Arundo. It belongs to the Poaceae family of the Arundina tribe. The genus includes also Arundo plinii, Arundo collina and Arundo mediterranea.

It is considered one of the largest herbaceous grasses as its height could reach more than 8 metres [77,108–110]. Several stems grow from the rhizome buds during all the vegetative season, forming dense clumps.A. donax stem is a hollow, segmented culm that measures from 1 to 4 centimeters in diameter and is able to branch during the second year of growth. Alternate leaves (5–8 cm wide and 30–70 cm long) are produced from the stem nodes, to which they are firmly wrapped [108, 111]. Stems and leaves are characterized by a relatively high content of silica, caused by the presence of siliceous cells associated with vascular bundles in the epidermal layer [108]. A. donax is reported to be a sterile species, and the propagation of this species is by agamic reproduction, occurring through regrowth of rhizome fragments and growth of shoots from stem nodes [112–114]. The adaptability to extreme soil conditions combined with rapid and vigorous growth makesA. donax an interesting subject for environmental studies on phytoremediation treatments. The use of plants to remove contaminants from polluted water and soil can be an advantageous strategy, which can also be used to remove metals that usually cannot be efficiently biodegraded. Studies indicate that A. donax may have a potential use for phytoremediation purposes. The plant is able to efficiently transfer arsenic absorbed from the growing medium and efficiently accumulate it into the shoots, showing a good tolerance to the presence of the metal [77].

Plant uptake and metabolism of arsenic has recently been reviewed by Tripathi et al. [115] and Zhao et al. [116]. Arsenic in the environment mainly exists in two inorganic oxidation states, arsenate (As (V)) and arsenite (As (III)). Phytoremediation of metals is the ability of plants to continually accumulate and detoxify metals in their system. In soil, arsenic exists in two forms, arsenate, thermodynamically stable under aerobic conditions, and arsenite under anaerobic conditions. As (V) is believed to interfere with oxidative phosphorylation, while As (III) may inhibit enzymatic activity by binding to thiol group. As (V) and As (III) enter plant cells via phosphate transporters and aqua glycophorins, respectively, as reviewed by Bhattacharjee et al. [117].

Once taken up, As (V) is reduced to As (III), catalyzed largely by arsenate reductases, members of the super family of protein tyrosine phosphatase (PTPase) [118] as shown in Figure 1. As (III) can then be complexed with glutathione (GSH) or phytochelatins (PCs). Raab et al. [103] identified up to 14 different species of arsenic complexes in sunflower plants. As (III) or complexed As (III) is then transported across the tonoplast and sequestered in the vacuole. Most data support the idea that arsenic is translocated from the roots to the tissues above ground, mostly in the form of As (III) [119, 120]. As (III) can be methylated to form monomethyl arsenate (MMAs (V)), dimethyl arsenate (DMAs (V)), and trimethyl arsine oxide (TMAO (V)) in plants [116,121]. Complexation of As (III) with PCs or GSH is an efficient way to detoxify arsenic, probably because the complexes are pumped and sequestered in the vacuole catalyzed by the homologues of multidrug resistance proteins (MRPs), members of the ABC super family [122].

Although these studies indicated the feasibility of over expressing phytochelator synthase (PCS) and/or -glutamylcysteine synthetase (-ECS) for increasing arsenic accumulation and concomitantly tolerance, there are no direct data on the site of arsenic storage in these transgenic lines; thus it remains unclear whether the complexed As (III) is primarily vacuolar or remains in the cytoplasm. It is possible that transport of complexed As (III) or even free As (III) across the tonoplast membrane is a potentially the rate-limiting step in overall arsenic tolerance and accumulation. Yet, to date, there are no reports of genetic engineering of tonoplast transport [123].

One of the key properties of arsenic hyperaccumulators such as Pteris vittata is a highly efficient system of arsenic translocation from root to shoot [119, 124], while most non-hyperaccumulators usually have a low mobility rate compared to P. vittata. Arsenic mobility from root to shoot varies considerably among different plant species, suggesting that it is under genetic control. A key step in arsenic translocation from root to shoot is arsenic loading to the xylem, a process that is not well understood. Ma et al. [125, 126] identified a gene encoding an efflux protein, Lsi2, which is responsible for arsenite loading into the xylem, as arsenite is the dominant arsenic species in the xylem. An Lsi2 mutation resulted in a nearly 50% reduction in arsenic accumulation in the shoot. Lsi2 is a homologue of the E. coli ArsB, which is an As (III)/H$^+$ exchanger that confers bacterial arsenite resistance [127]. The plant efflux protein apparently transports both metalloids As (III) and Si (IV). Methylated arsenic species have been detected in several plant species, including rice grain [128, 129], and recent data suggest that this is the result of endogenous methylation by the plants themselves [121]. The final product of the methylation pathway is the gas trimethyl arsine (TMAs (III)), which can be volatilized from the plant [78].

Disposal of Waste

The contaminated plant biomass can be digested or ashed to reduce its volume (95%), and the resulting small volume of material can be processed as an "ore" to recover the contaminant (e.g., valuable heavy metals and radionuclides). If recycling the metal is not economically feasible, the relatively small amount of ash (compared to the original biomass or the extremely large volume of contaminated soil) can be disposed of in an appropriate manner [51]. Various disposal options have been presented in Figure 2.

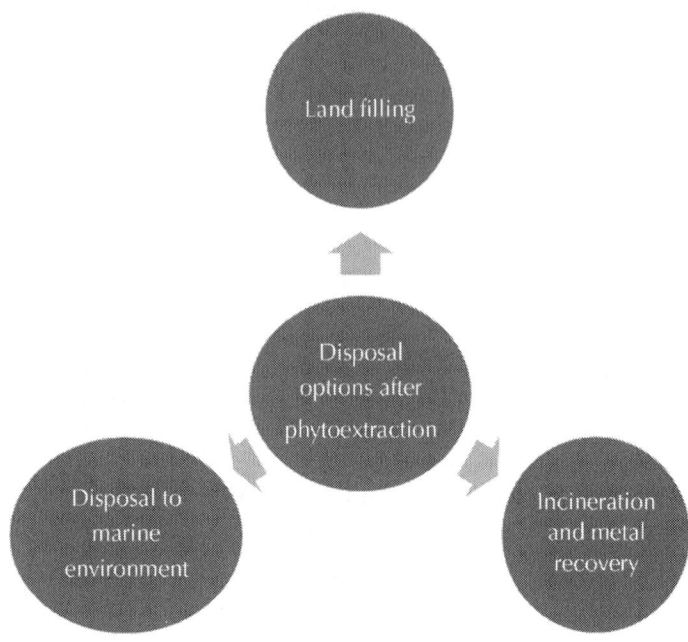

Figure 2: Various options for arsenic containing wastes disposal.

Marine systems also have a particular ability to biotransform and detoxify inorganic arsenic, presumably due to their evolution in an arsenic-containing environment; seawater contains approximately 1 $\mu gAsL^{-1}$ primarily as arsenate [130]. Arsenite added to seawater is likely to be quickly converted to arsenate, the more thermodynamically stable form [131]. The first stage of this detoxification process is formation of arsenosugars by algae; perhaps other organisms may also be involved and the final arsenic metabolite appears to be arsenobetaine, a stable nontoxic form of arsenic found in all marine animals [130]. Some marine unicellular algae can carry out this biotransformation at arsenate concentrations 1000-fold ambient levels (i.e. at 1000 $\mu gAsL^{-1}$) [132]. Thus, it may be possible to dispose of the high arsenic fern directly into the open sea where it would degrade (contributing marginally to nutrient levels) and release inorganic arsenic which could be converted to nontoxic forms by natural processes. There may be possible ecological effects from the initial increased arsenic concentrations in seawater such as species changes in algae populations [133, 134].

CONCLUSIONS

Arsenic toxicity in soil and water is an increasing menace across the world and it is causing significant health damage to people living in developing and third world countries. It can be declared as a global hazard. Such a situation demands low-cost, technologically simple and point of use solutions to arsenic toxicity.

Phytoremediation is a sustainable option for developing countries which are hit by economic crisis and thus cannot afford technologically sophisticated solutions for their huge populations. Many plant species especially aquatic macrophytes and some wetland plants have shown promising ability to uptake arsenic from contaminated environments. Free metal ions in the soil solution are absorbed by plants and are reduced as metal chelates using specific high-affinity ligands (like oxygen-donor ligands, sulfur-donor ligands, and nitrogen-donor ligands). Bioaccumulation in stems and leaves along phytovolatilization have been shown to be possible tolerance mechanisms by plants against arsenic contamination.

REFERENCES

1.   M. Pantsar-Kallio and A. Korpela, "Analysis of gaseous arsenic species and stability studies of arsine and trimethylarsine by gas chromatography-mass spectrometry," Analytica Chimica Acta, vol. 410, no. 1-2, pp. 65–70, 2000.

2.   C. F. Balasoiu, G. J. Zagury, and L. Deschênes, "Partitioning and speciation of chromium, copper, and arsenic in CCA-contaminated soils: influence of soil composition," Science of the Total Environment, vol. 280, no. 1–3, pp. 239–255, 2001.

3.   B. Rathinasabapathi and L. Q. Ma, "Arsenic hyperaccumulating ferns and their application to phytoremediation of arsenic contaminated sites," in Floriculture, Ornamental and Plant Biotechnology: Advances and Topical Issues, pp. 304–311, Global Science Books, London, UK, 1st edition, 2006.

4.   Z. Feng, Y. Xia, D. Tian et al., "DNA damage in buccal epithelial cells from individuals chronically exposed to arsenic via drinking

water in Inner Mongolia, China," Anticancer Research, vol. 21, no. 1, pp. 51–57, 2001.

5. R. D. Tripathi, S. Srivastava, S. Mishra et al., "Arsenic hazards: strategies for tolerance and remediation by plants," Trends in Biotechnology, vol. 25, no. 4, pp. 158–165, 2007.

6. R. Zhao, M. Zhao, H. Wang, Y. Taneike, and X. Zhang, "Arsenic speciation in moso bamboo shoot—a terrestrial plant that contains organoarsenic species," Science of the Total Environment, vol. 371, no. 1–3, pp. 293–303, 2006.

7. SOS Arsenic Poisoning in Bangladesh/India, 2011, http://www.sos-arsenic.net.

8. C.-J. Chen, C. W. Chen, M.-M. Wu, and T.-L. Kuo, "Cancer potential in liver, lung, bladder and kidney due to ingested inorganic arsenic in drinking water," British Journal of Cancer, vol. 66, no. 5, pp. 888–892, 1992.

9. S. Kapaj, H. Peterson, K. Liber, and P. Bhattacharya, "Human health effects from chronic arsenic poisoning—a review," Journal of Environmental Science and Health A, vol. 41, no. 10, pp. 2399–2428, 2006.

10. Committee on Medical, Biological Effects of Environmental Pollutants, and National Research Council,Arsenic: Medical and Biological Effects of Environmental Pollutants, National Academy of Sciences, Washington, DC, USA, 1977.

11. R. T. Nickson, J. M. McArthur, B. Shrestha, T. O. Kyaw-Myint, and D. Lowry, "Arsenic and other drinking water quality issues, Muzaffargarh District, Pakistan," Applied Geochemistry, vol. 20, no. 1, pp. 55–68, 2005.

12. P. C. Nagajyoti, K. D. Lee, and T. V. M. Sreekanth, "Heavy metals, occurrence and toxicity for plants: a review," Environmental Chemistry Letters, vol. 8, no. 3, pp. 199–216, 2010.

13. National Research Council (NRC), Current Issues and Studies, 1977, www.eric.ed.gov/ERICWebPortal.

14. B. K. Mandal and K. T. Suzuki, "Arsenic round the world: a review," Talanta, vol. 58, no. 1, pp. 201–235, 2002.

15. W. J. Fitz and W. W. Wenzel, "Arsenic transformations in the soil-rhizosphere-plant system: fundamentals and potential application

to phytoremediation," Journal of Biotechnology, vol. 99, no. 3, pp. 259–278, 2002.

16. B. K. Chopra, S. Bhat, I. P. Mikheenko et al., "The characteristics of rhizosphere microbes associated with plants in arsenic-contaminated soils from cattle dip sites," Science of the Total Environment, vol. 378, no. 3, pp. 331–342, 2007.

17. X. Sun and H. E. Doner, "An investigation of arsenate and arsenite bonding structures on goethite by FTIR," Soil Science, vol. 161, no. 12, pp. 865–872, 1996.

18. R. R. Frost and R. A. Griffin, "Effect of pH on adsorption of as and selenium from landfill leachate by clay minerals," Soil Science Society of America Journal, vol. 41, no. 1, pp. 53–57, 1977.

19. S. H. Mueller, R. J. Goldfarb, G. L. Farmer, R. Sanzolone, M. Adams, and P. Theodorakus, "A seasonal study of arsenic and groundwater geochemistry in Fairbanks, Alaska," in Proceedings of the USGS Workshop on Arsenic in the Environment, Denver, Colorado, February 2001.

20. N. P. Nikolaidis, G. M. Dobbs, J. Chen, and J. A. Lackovic, "Arsenic mobility in contaminated lake sediments," Environmental Pollution, vol. 129, no. 3, pp. 479–487, 2004.

21. P. L. Smedley, H. B. Nicolli, D. M. J. Macdonald, A. J. Barros, and J. O. Tullio, "Hydrogeochemistry of arsenic and other inorganic constituents in groundwaters from La Pampa, Argentina," Applied Geochemistry, vol. 17, no. 3, pp. 259–284, 2002.

22. D. C. Adriano, Trace Elements in Terrestrial Environments: Bio Geochemistry, Bioavailability, and Risk of Metals, Springer, New York, NY, USA, 2nd edition, 2001.

23. J. C. Ng, J. Wang, and A. Shraim, "A global health problem caused by arsenic from natural sources," Chemosphere, vol. 52, no. 9, pp. 1353–1359, 2003.

24. E. K. Porter and P. J. Peterson, "Arsenic accumulation by plants on mine waste (United Kingdom)," Science of the Total Environment, vol. 4, no. 4, pp. 365–371, 1975.

25. T. de Koe, Arsenic resistance in submeditemanean Agrotis species, [Ph.D. thesis], Vrije Universiteit, Amsterdam, The Netherlands, 1994.

26. A. M. Benavides, The effect of arsenic speciation on arsenic uptake and fate in the presence of the hyper-accumulating species Pteris cretica, [Master of Science], 2007.

27. S. Mahimairaja, N. S. Bolan, D. C. Adriano, and B. Robinson, "Arsenic contamination and its risk management in complex environmental settings," Advances in Agronomy, vol. 86, pp. 1–81, 2005.

28. X. Y. Zhang, F. F. Lin, M. T. F. Wong, X. L. Feng, and K. Wang, "Identification of soil heavy metal sources from anthropogenic activities and pollution assessment of Fuyang County, China,"Environmental Monitoring and Assessment, vol. 154, no. 1–4, pp. 439–449, 2009.

29. A. O. Fayiga, Phytoremediation of arsenic-contaminated soil and groundwater, [Dissertation for the Degree of Doctor of Philosophy], University of Florida, 2005.

30. W. J. Mwegoha, "The use of phytoremediation technology for batement soil and groundwater pollute on in Tanzania: opportunities and challenges," Journal of Sustainable Development in Africa, vol. 10, pp. 140–156, 2008.

31. J. O. Nriagu and J. M. Pacyna, "Quantitative assessment of worldwide contamination of air, water and soils by trace metals," Nature, vol. 333, no. 6169, pp. 134–139, 1988.

32. R. R. Rodriguez, N. T. Basta, S. W. Casteel, F. P. Armstrong, and D. C. Ward, "Chemical extraction methods to assess bioavailable arsenic in soil and solid media," Journal of Environmental Quality, vol. 32, no. 3, pp. 876–884, 2003.

33. E. Smith, R. Naidu, and A. M. Alston, "Chemistry of inorganic arsenic in soils: II. Effect of phosphorus, sodium, and calcium on arsenic sorption," Journal of Environmental Quality, vol. 31, no. 2, pp. 557–563, 2002.

34. A. Davis, D. Sherwin, R. Ditmars, and K. A. Hoenke, "An analysis of soil arsenic records of decision,"Environmental Science and Technology, vol. 35, no. 12, pp. 2401–2406, 2001.

35. R. J. Ampiah-Bonney, J. F. Tyson, and G. R. Lanza, "Phytoextraction of arsenic from soil by Leersia oryzoides," International Journal of Phytoremediation, vol. 9, no. 1, pp. 31–40, 2007.

36. A. C. Aurilio, J. L. Durant, H. F. Hemond, and M. L. Knox, "Sources and distribution of arsenic in the Aberjona Watershed, Eastern Massachusetts," Water, Air, and Soil Pollution, vol. 81, no. 3-4, pp. 265–282, 1995.

37. A. L. Salido, K. L. Hasty, J.-M. Lim, and D. J. Butcher, "Phytoremediation of arsenic and lead in contaminated soil using Chinese Brake ferns (Pteris vittata) and Indian mustard (Brassica juncea),"International Journal of Phytoremediation, vol. 5, no. 2, pp. 89–103, 2003.

38. Environment Canada, Canadian Soil Quality Guidelines for Arsenic: Environmental and Human Health, Science Policy and Environmental Quality Branch, Guidelines Division, Ottawa, Canada, 1996.

39. H. K. Das, A. K. Mitra, P. K. Sengupta, A. Hossain, F. Islam, and G. H. Rabbani, "Arsenic concentrations in rice, vegetables, and fish in Bangladesh: a preliminary study," Environment International, vol. 30, no. 3, pp. 383–387, 2004.

40. T. Roychowdhury, H. Tokunaga, and M. Ando, "Survey of arsenic and other heavy metals in food composites and drinking water and estimation of dietary intake by the villagers from an arsenic-affected area of West Bengal, India," Science of the Total Environment, vol. 308, no. 1–3, pp. 15–35, 2003.

41. A. K. Sharma, Arsenic removal from water using naturally occurring Iron, and the associated benefits on health in affected regions. [Ph.D. thesis], Institute of Environment & Resources Technical University of Denmark, 2006.

42. R. Khan, S. H. Israili, H. Ahmad, and A. Mohan, "Heavy metal pollution assessment in surface water bodies and its suitability for irrigation around the Neyevli lignite mines and associated industrial complex, Tamil Nadu, India," Mine Water and the Environment, vol. 24, no. 3, pp. 155–161, 2005.

43. W. Rehman, A. Zeb, N. Noor, and M. Nawaz, "Heavy metal pollution assessment in various industries of Pakistan," Environmental Geology, vol. 55, no. 2, pp. 353–358, 2008.

44. A. Malik, "Environmental challenge vis a vis opportunity: the case of water hyacinth," Environment International, vol. 33, no. 1, pp. 122–138, 2007.

45. United Nations Environment Program (UNEP), Global Environment Outlook 2000, Earthscan, UK, 1999.

46. WHO, Guidelines for Drinking-Water Quality: Recommendations, vol. 1, World Health Organisation, Geneva, Switzerland, 2nd edition, 1993.

47. R. T. Nickson, J. M. McArthur, B. Shrestha, T. O. Kyaw-Myint, and D. Lowry, "Arsenic and other drinking water quality issues, Muzaffargarh District, Pakistan," Applied Geochemistry, vol. 20, no. 1, pp. 55–68, 2005.

48. B. Shrestha, "Drinking water quality: future directions for UNICEF in Pakistan," Consultancy Report 2 of 3, Water Quality, SWEET Project, UNICEF, Pakistan, Islamabad, 2002.

49. Z. L. He, X. E. Yang, and P. J. Stoffella, "Trace elements in agroecosystems and impacts on the environment," Journal of Trace Elements in Medicine and Biology, vol. 19, no. 2-3, pp. 125–140, 2005.·

50. A. H. Malik, Z. M. Khan, Q. Mahmood, S. Nasreen, and Z. A. Bhatti, "Perspectives of low cost arsenic remediation of drinking water in Pakistan and other countries," Journal of Hazardous Materials, vol. 168, no. 1, pp. 1–12, 2009.

51. B. R. Bondada and L. Q. Ma, "Tolerance of heavy metals in vascular plants: arsenic hyperaccumulation by chinese, brake fern (Pterzs vzttata L.)," in Pteridology in the New Millennium, S. Chandra and M. Srivastava, Eds., pp. 397–420, 2003.

52. M. M. Lasat, "Phytoextraction of metals from contaminated soil: a review of plant, soil, metal interaction and assessment of pertinent agronomic issues," Journal of Hazardous Substance Research, vol. 2, pp. 1–25, 2000.

53. A. Allen, "Containment landfills: the myth of sustainability," Engineering Geology, vol. 60, no. 1–4, pp. 3–19, 2001.

54. T. Mahmood, S. A. Malik, Z. Hussain, I. Qamar, and H. A. Mateen, A Review of Phytoremediation Technology for Contaminated Soil and Water, ESDev. CIIT, Abbottabad, Pakistan, 2007.

55. M. Santiago and N. S. Bolan, "Phytoremediation of arsenic contaminated soil and water," in Proceedings of the 19th World Congress of Soil Science. Soil Solutions for a Changing World, Brisbane, Australia, August 2010.

56. A. A. Meharg, "Integrated tolerance mechanisms: constitutive and adaptive plant responses to elevated metal concentrations in the environment," Plant, Cell and Environment, vol. 17, no. 9, pp. 989–993, 1994.

57. Z. Yanqun, L. Yuan, C. Schvartz, L. Langlade, and L. Fan, "Accumulation of Pb, Cd, Cu and Zn in plants and hyperaccumulator choice in Lanping lead-zinc mine area, China," Environment International, vol. 30, no. 4, pp. 567–576, 2004.

58. K. G. Anil, M. Yunus, and P. K. Pandey, "Bioremediation: ecotechnology for the present century,"International Society of Environmental Botanists, vol. 9, no. 2, 2003.

59. S. E. Rocovich and D. A. West, "Arsenic tolerance in a population of the grass Andropogon scopariusMichx," Science, vol. 188, no. 4185, pp. 263–264, 1975.

60. A. J. Pollard, "Diversity of metal tolerance in Plantago lanceolara L. from the Southeastern United States," New Phytologist, vol. 86, pp. 109–117, 1980.

61. A. A. Mehrag and M. R. Macnair, "An altered phosphate uptake system in arsenate-tolerant Holcus lantus L.," New Phytologist, vol. 116, no. 1, pp. 29–35, 1990.

62. J. M. Wells and D. H. S. Richardson, "Anion accumulation by the moss Hylocomium splendens: uptake and competition studies involving arsenate, selenate, selenite, phosphate, sulphate and sulphite," New Phytologist, vol. 101, no. 4, pp. 571–583, 1985.

63. E. Nieboer, D. Padovan, P. Lavoie, and D. H. S. Richardson, "Anion accumulation by lichens. II. Competition and toxicity studies involving arsenate, phosphate, sulphate and sulphite," New Phytologist, vol. 96, no. 1, pp. 83–93, 1984.

64. R. E. Beever and D. J. W. Burns, "Phosphorus uptake, storage and utilization by fungi," Advances in Botanical Research, vol. 8, pp. 127–219, 1981.

65. S. Silver and T. K. Misra, "Plasmid-mediated heavy metal resistances," Annual Review of Microbiology, vol. 42, pp. 717–743, 1988.

66. S. Monni, H. Bücking, and I. Kottke, "Ultrastructural element localization by EDXS in Empetrum nigrum," Micron, vol. 33, no. 4, pp. 339–351, 2002.

67. W. Tossellr, K. Binard, and M. T. Rafferty, "Uptake of arsenic by tamarisk and eucalyptus under saline conditions," in Bioremediation and Phyroremediarion of Chlorinated and Recalcitrant Compounds, G. B. Wickramanayake, A. R. Gavaskar, B. C. Meman, and V. C. Magar, Eds., pp. 485–492, Battelle Press, Columbus, Richmond, 2000.

68. L. Q. Ma, K. M. Komar, C. Tu, W. Zhang, Y. Cai, and E. D. Kennelley, "A fern that hyperaccumulates arsenic," Nature, vol. 409, article 579, no. 6836, 2001.

69. A. J. M. Baker, R. Brooks, and R. Reeves, "Growing for gold and copper and zinc," New Scientist, vol. 10, no. 1603, pp. 44–48, 1988.

70. M. R. Macnair, "The evolution of plants in metal-contaminated environments," in Environmental Stress, Adaptation and Evolution, R. Bijlsma and V. Loeschcke, Eds., pp. 3–24, Birkhauser, Boston, Mass, USA, 1997.

71. A. A. Meharg and M. R. Macnair, "Suppression of the high affinity phosphate uptake system: a mechanism of arsenate tolerance in Holcus lanatus L," Journal of Experimental Botany, vol. 43, no. 4, pp. 519–524, 1992.

72. S. M. Ross and K. J. Kae, "The meaning of metal toxicity in soil-plant systems," in Toxic Metals in Soil-Plant Systems, S. M. Ross, Ed., pp. 153–188, John Wiley & Sons, New York, NY, USA, 1994.

73. M. R. Macnair and Q. Cumbes, "Evidence that arsenic tolerance in Holcus lanatus L. is caused by an altered phosphate uptake system," New Phytologist, vol. 107, no. 2, pp. 387–394, 1987.

74. A. J. Asher and O. F. Reay, "Arsenic uptake by barley seedlings," Australian Journal of Plant Physiology, vol. 6, no. 4, pp. 459–466, 1979.

75. A. I. Ullrich-eberius, A. Sanz, and A. J. Novacky, "Evaluation of arsenate- and vanadate-associated changes of electrical membrane potential and phosphate transport in Lemna gibba G1," Journal of Experimental Botany, vol. 40, no. 1, pp. 119–128, 1989.

76. D. J. Thompson, "A chemical hypothesis for arsenic methylation in mammals," Chemico-Biological Interactions, vol. 88, no. 2-3, pp. 89–114, 1993.

77. N. Mirza, Q. Mahmood, A. Pervez et al., "Phytoremediation potential of Arundo donax in arsenic-contaminated synthetic wastewater," Bioresource Technology, vol. 101, no. 15, pp. 5815–5819, 2010.·

78. N. Mirza, A. Pervez, Q. Mahmood, and S. S. Ahmad, "Phytoremediation of arsenic (As) and mercury (Hg) contaminated soil," World Applied Sciences Journal, vol. 8, pp. 113–118, 2010.

79. F. E. C. Sneller, L. M. van Heerwaarden, F. J. L. Kraaijeveld-Smit et al., "Toxicity of arsenate in Silene vulgaris, accumulation and degradation of arsenate-induced phytochelatins," New Phytologist, vol. 144, no. 2, pp. 223–232, 1999.

80. A. R. Marin, S. R. Pezashkip, H. Masschelen, and H. S. Choi, "Effect of dimethylarsenic acid on growth, tissue arsenic, and photosynthesis of rice plants," Journal of Plant Nutrition, vol. 16, pp. 865–880, 2002.

81. I. J. Pickering, R. C. Prince, M. J. George, R. D. Smith, G. N. George, and D. E. Salt, "Reduction and coordination of arsenic in Indian mustard," Plant Physiology, vol. 122, no. 4, pp. 1171–1177, 2000.

82. J. Hartley-Whitaker, G. Ainsworth, R. Vooijs, W. Ten Bookum, H. Schat, and A. A. Meharg, "Phytochelatins are involved in differential arsenate tolerance in Holcus lanatus," Plant Physiology, vol. 126, no. 1, pp. 299–306, 2001.

83. A. J. M. Baker, S. P. McGrath, R. D. Reeves, and J. A. C. Smith, "Metal hyperaccumulator plants: a review of the ecology and physiology of a biological resource for phytoremediation of metal polluted soils," in Phytoremediation of Contaminated Soil and Water, N. Terry and G. Banuelos, Eds., pp. 5–107, Lewis, Boca Raton, Fla, USA, 2000.

84. J. F. Ma, S. Hiradate, K. Nomoto, T. Iwashita, and H. Matsumoto, "Internal detoxification mechanism of Al in hydrangea: Identification of Al form in the leaves," Plant Physiology, vol. 113, no. 4, pp. 1033–1039, 1997.

85. R. Gabbrielli, P. Gremigni, L. B. Morassi, T. Pandolfini, and P. Medeghini, "Some aspects of ni tolerance in slyssum bertolonii Desv.: strategies of metal distribution and accumulation," in Ecologie des Milieux sur Roches Ultramafique et sur Sols MEtalliferes, T. Jaffre, R. D. Reeves, and T. Becquer, Eds., pp. 225–

227, Documents scientifiques et techniques ORSTOM, Noumea New, Caledonia, 1997.

86.   F. A. Homer, R. D. Reeves, and R. R. Brooks, "The possible involvement of amino acids in nickel chelation in some nickel-accumulating plants," in Current Opinion in Phytochemistry, vol. 14, pp. 31–37, 1995.

87.   N. J. Robinson, A. M. Tommey, C. Kuske, and P. J. Jackson, "Plant metallothioneins," Biochemical Journal, vol. 295, no. 1, pp. 1–10, 1993.

88.   H. Tohoyama, M. Inouhe, M. Joho, and T. Murayama, "Production of metallothionein in copper- and cadmium-resistant strains of Saccharomyces cerevisiae," Journal of Industrial Microbiology, vol. 14, no. 2, pp. 126–131, 1995.

89.   M. R. Macnair, "The evolution of plants in metal contaminated environments," in Environmental Stress, Adaptation and Evolution, R. Bijlsma and V. Loeschcke, Eds., pp. 3–24, Birkhauser, Boston, Mass, USA, 1997.

90.   D. E. Salt, R. D. Smith, and I. Raskin, "Phytoremediation," Annual Review of Plant Biology, vol. 49, pp. 643–668, 1998.

91.   N. Mirza, A. Pervez, Q. Mahmood, M. M. Shah, and M. N. Shafqat, "Ecological restoration of arsenic contaminated soil by Arundo donax L.," Ecological Engineering, vol. 37, no. 12, pp. 1949–1956, 2011.·

92.   M. Doucleff and N. Terry, "Pumping out the arsenic," Nature Biotechnology, vol. 20, no. 11, pp. 1094–1095, 2002.

93.   W. E. Rauser, "Phytochelatins and related peptides. Structure, biosynthesis, and function," Plant Physiology, vol. 109, no. 4, pp. 1141–1149, 1995.

94.   M. E. V. Schmoger, M. Oven, and E. Grill, "Detoxification of arsenic by phytochelatins in plants," Plant Physiology, vol. 122, no. 3, pp. 793–801, 2000.

95.   F. J. Zhao, J. R. Wang, J. H. A. Barker, H. Schat, P. M. Bleeker, and S. P. McGrath, "The role of phytochelatins in arsenic tolerance in the hyperaccumulator Pteris vittata," New Phytologist, vol. 159, no. 2, pp. 403–410, 2003.

96.   T. Maitani, H. Kubota, K. Sato, and T. Yamada, "The composition of metals bound to class III metallothionein (phytochelatin and

its desglycyl peptide) induced by various metals in root cultures of Rubia tinctorum," Plant Physiology, vol. 110, no. 4, pp. 1145–1150, 1996.

97. E. Grill, E. L. Winnacker, and M. H. Zenk, "Phytochelatins: the principal heavy-metal complexing peptides of higher plants," Science, vol. 230, no. 4726, pp. 674–676, 1985.

98. E. Grill, S. Loffler, E. L. Winnacker, and M. H. Zenk, "Phytochelatins, the heavy metal binding peptides of plants are synthesized from glutathione by a specific y-glutamylcysteine dipeptidyl transpeptidase (phytochelatin synthase)," Proceedings of the National Academy of Sciences of the USA, vol. 86, no. 18, pp. 6838–6842, 1989.

99. E. Grill, E.-L. Winnacker, and M. H. Zenk, "Phytochelatins, a class of heavy-metal-binding peptides from plants, are functionally analogous to metallothioneins," Proceedings of the National Academy of Sciences of the United States of America, vol. 84, no. 2, pp. 439–443, 1987.

100. R. Vögeli-Lange and G. J. Wagner, "Subcellular localization of cadmium and cadmium-binding peptides in tobacco leaves: Implication of a transport function for cadmium-binding peptides," Plant Physiology, vol. 92, no. 4, pp. 1086–1093, 1990.

101. C. Cobbett and P. Goldsbrough, "Phytochelatins and metallothioneins: Roles in heavy metal detoxification and homeostasis," Annual Review of Plant Biology, vol. 53, pp. 159–182, 2002.

102. W. Chintakovid, P. Visoottiviseth, S. Khokiattiwong, and S. Lauengsuchonkul, "Potential of the hybrid marigolds for arsenic phytoremediation and income generation of remediators in Ron Phibun District, Thailand," Chemosphere, vol. 70, no. 8, pp. 1532–1537, 2008.

103. A. Raab, H. Schat, A. A. Meharg, and J. Feldmann, "Uptake, translocation and transformation of arsenate and arsenite in sunflower (Helianthus annuus): Formation of arsenic-phytochelatin complexes during exposure to high arsenic concentrations," New Phytologist, vol. 168, no. 3, pp. 551–558, 2005.·

104. X. Wang, Y. Wang, Q. Mahmood et al., "The effect of EDDS addition on the phytoextraction efficiency from Pb contaminated

soil by Sedum alfredii Hance," Journal of Hazardous Materials, vol. 168, no. 1, pp. 530–535, 2009.

105. W. Zhang, Y. Cai, C. Tu, and L. Q. Ma, "Arsenic speciation and distribution in an arsenic hyperaccumulating plant," Science of the Total Environment, vol. 300, no. 1-3, pp. 167–177, 2002.

106. G. P. Bell, "Ecology and management of Arundo donax, and approaches to riparian habitat restoration in southern California," in Plant Invasions: Studies From North America and Europe, J. H. Brock, M. Wade, P. Pysek, and D. Green, Eds., pp. 103–113, Blackhuys, Leiden, The Netherlands, 1997.

107. T. L. Dudley, "Arundo donax L.," in Invasive Plants of California›s Wildlands, C. C. Brossard, J. M. Randall, and M. C. Hoshovsky, Eds., pp. 53–58, University of California Press, Berkeley, Calif, USA, 2000.

108. R. E. Perdue, "Arundo donax-Source of musical reeds and industrial cellulose," Economic Botany, vol. 12, no. 4, pp. 368–404, 1958.

109. D. F. Spencer, P.-S. Liow, W. K. Chan, G. G. Ksander, and K. D. Getsinger, "Estimating Arundo donax shoot biomass," Aquatic Botany, vol. 84, no. 3, pp. 272–276, 2006.

110. P. R. Frandsen, "Team arundo: interagency cooperation to control giant reed cane (Arundo donax)," inAssessment and Management of Plant Invasions, O. L. James and J. W. Thiert, Eds., pp. 244–247, Springer, New York, NY, USA, 1997.

111. G. C. Tucker, "The genera of Arundinoideae (Graminae) in the southeastern United States," Journal of the Arnold Arboretum, vol. 71, pp. 145–177, 1990.

112. J. M. Boland, "The importance of layering in the rapid spread of Arundo donax (giant reed)," Madroño, vol. 53, no. 4, pp. 303–312, 2006.

113. A. B. Boose and J. S. Holt, "Environmental effects on asexual reproduction in Arundo donax," Weed Research, vol. 39, no. 2, pp. 117–127, 1999.

114. J. M. DiTomaso and E. A. Healey, Aquatic and Riparian Weeds of the West, vol. 3421 of Agriculture and Natural Resources, University of California, 2003.

115. R. D. Tripathi, S. Srivastava, S. Mishra et al., "Arsenic hazards: strategies for tolerance and remediation by plants," Trends in Biotechnology, vol. 25, no. 4, pp. 158–165, 2007.

116. F. J. Zhao, J. F. Ma, A. A. Meharg, and S. P. McGrath, "Arsenic uptake and metabolism in plants," New Phytologist, vol. 181, no. 4, pp. 777–794, 2009.

117. H. Bhattacharjee, R. Mukhopadhyay, S. Thiyagarajan, and B. P. Rosen, "Aquaglyceroporins: ancient channels for metalloids," Journal of Biology, vol. 7, no. 9, article 33, 2008.

118. R. Mukhopadhyay and B. P. Rosen, "Arsenate reductases in prokaryotes and eukaryotes," Environmental Health Perspectives, vol. 110, no. 5, pp. 745–748, 2002.

119. X. Y. Xu, S. P. McGrath, and F. J. Zhao, "Rapid reduction of arsenate in the medium mediated by plant roots," New Phytologist, vol. 176, no. 3, pp. 590–599, 2007.

120. Y. H. Su, S. P. McGrath, Y. G. Zhu, and F. J. Zhao, "Highly efficient xylem transport of arsenite in the arsenic hyperaccumulator Pteris vittata," New Phytologist, vol. 180, no. 2, pp. 434–441, 2008.

121. J. Wu, R. Zhang, and R. M. Lilley, "Methylation of arsenic in vitro by cell extracts from bentgrass (Agrostis tenuis): effect of acute exposure of plants to arsenate," Functional Plant Biology, vol. 29, no. 1, pp. 73–80, 2002.

122. R. Tommasini, E. Vogt, M. Fromenteau et al., "An ABC-transporter of Arabidopsis thaliana has both glutathione-conjugate and chlorophyll catabolite transport activity," Plant Journal, vol. 13, no. 6, pp. 773–780, 1998.

123. Y.-G. Zhu and B. P. Rosen, "Perspectives for genetic engineering for the phytoremediation of arsenic-contaminated environments: from imagination to reality?" Current Opinion in Biotechnology, vol. 20, no. 2, pp. 220–224, 2009.

124. G.-L. Duan, Y.-G. Zhu, Y.-P. Tong, C. Cai, and R. Kneer, "Characterization of arsenate reductase in the extract of roots and fronds of Chinese brake fern, an arsenic hyperaccumulator," Plant Physiology, vol. 138, no. 1, pp. 461–469, 2005.

125. J. F. Ma, K. Tamai, M. Ichii, and G. F. Wu, "A rice mutant defective in Si uptake," Plant Physiology, vol. 130, no. 4, pp. 2111–2117, 2002.

126. F. M. Jian, K. Tamai, N. Yamaji et al., "A silicon transporter in rice," Nature, vol. 440, no. 7084, pp. 688–691, 2006.

127. Y.-L. Meng, Z. Liu, and B. P. Rosen, "As(III) and Sb(III) uptake by GlpF and efflux by ArsB inEscherichia coli," Journal of Biological Chemistry, vol. 279, no. 18, pp. 18334–18341, 2004.

128. P. N. Williams, A. H. Price, A. Raab, S. A. Hossain, J. Feldmann, and A. A. Meharg, "Variation in arsenic speciation and concentration in paddy rice related to dietary exposure," Environmental Science and Technology, vol. 39, no. 15, pp. 5531–5540, 2005.

129. Y.-G. Zhu, G.-X. Sun, M. Lei et al., "High percentage inorganic arsenic content of mining impacted and nonimpacted chinese rice," Environmental Science and Technology, vol. 42, no. 13, pp. 5008–5013, 2008.·

130. K. Francesconi, P. Visoottiviseth, W. Sridokchan, and W. Goessler, "Arsenic species in an arsenic hyperaccumulating fern, Pityrogramma calomelanos: a potential phytoremediator of arsenic-contaminated soils," Science of the Total Environment, vol. 284, no. 1–3, pp. 27–35, 2002.

131. W. R. Cullen and K. J. Reimer, "Arsenic speciation in the environment," Chemical Reviews, vol. 89, no. 4, pp. 713–764, 1989.

132. J. S. Edmonds, Y. Shibata, K. A. Francesconi, R. J. Rippingale, and M. Morita, "Arsenic transformations in short marine food chains studied by HPLC-ICP MS," Applied Organometallic Chemistry, vol. 11, no. 4, pp. 281–287, 1997.

133. J. G. Sanders and S. J. Cibik, "Adaptive behaviour of euryhaline phytoplankton communities to arsenic stress," in Marine Ecology Progress Series, vol. 22, pp. 199–205, 1985.

134. K. A. Francesconi and J. S. Edmonds, "Arsenic and Marine Organisms," Advances in Inorganic Chemistry, vol. 44, pp. 147–189, 1996.

Chapter 2

Experimental Approaches for Evaluating the Invasion Risk of Biofuel Crops

S Luke Flory[1], Kimberly A Lorentz[2], Doria R Gordon[3], and Lynn E Sollenberger[1]

Department, University of Florida, Gainesville, FL 32611, USA

[2]Forest Resources and Conservation, University of Florida, Gainesville, FL 32611, USA

[3]The Nature Conservancy, Department of Biology, University of Florida, Gainesville, FL 32611, USA

ABSTRACT

There is growing concern that non-native plants cultivated for bioenergy production might escape and result in harmful invasions in natural areas. Literature-derived assessment tools used to evaluate invasion risk are beneficial for screening, but cannot be used to assess novel cultivars or genotypes. Experimental approaches are needed to help quantify invasion risk but protocols for such tools are lacking.

We review current methods for evaluating invasion risk and make recommendations for incremental tests from small-scale experiments to widespread, controlled introductions. First, local experiments should be performed to identify conditions that are favorable for germination, survival, and growth of candidate biofuel crops. Subsequently, experimental introductions in semi-natural areas can be used to assess factors important for establishment and performance such as disturbance, founder population size, and timing of introduction across variable habitats. Finally, to fully characterize invasion risk, experimental introductions should be conducted across the expected geographic range of cultivation over multiple years. Any field-based testing should be accompanied by safeguards and monitoring for early detection of spread. Despite the costs of conducting experimental tests of invasion risk, empirical screening will greatly improve our ability to determine if the benefits of a proposed biofuel species outweigh the projected risks of invasions.

INTRODUCTION

Rising energy prices and the increasingly recognized role of fossil fuel consumption in climate change have driven the recent development of second-generation biofuels. However, concern is growing over the possibility of unintended environmental impacts from the hasty adoption of novel biofuel feedstock crops (Demirbas 2009). One of the primary concerns is that introduced biofuel crops will result in escapes into natural areas and widespread, ecologically damaging invasions (Raghu et al 2006, Richardson and Blanchard 2011). This concern is driven by the fact that many of the traits that maximize biofuel crop biomass yield are synonymous with the ecological traits of successful invasive species. Such traits include rapid accumulation of biomass, short generative time, perennial growth form, disease and pest resistance, and tolerance to drought, low soil fertility, and saline conditions (Raghu et al 2006, Barney and DiTomaso 2008, 2010, NISC 2009).

The desirability of these traits in bioenergy crops is apparent in the proposed cultivation of several fast growing species with a long history of invasive spread. Biofuel feedstock candidates such as giant reed (*Arundo donax*), miscanthus (*Miscanthus sinensis*), reed canary grass (*Phalaris arundinacea*), and Chinese tallow (*Triadica sebifera*) commonly appear on invasive species lists, thus their development as biofuel crops could clash with current efforts to manage these species (Davis *et al* 2010). In addition, selecting for certain traits in biofuel crop cultivars may increase the risk of biofuels becoming invasive. For example, while the sterility of *Miscanthus* × *giganteus* (triploid hybrid between the invasive *M. sinensis* and *M. saccharifloris*) has reduced concerns about invasiveness of this bioenergy crop (Barney and DiTomaso 2008), new efforts have produced a cultivar with viable seed that will be less expensive to propagate (Ross 2011). Furthermore, invasion risk for biofuel crops may be heightened by the sustained propagule pressure associated with large-scale cultivation (Davis *et al* 2011, Mack 2008, Minton and Mack 2010). The amount of land required in order for second-generation biofuel crops to meet biofuel production targets is expected to be in the millions of acres (Dale *et al* 2011), resulting in high potential for production and release of propagules. Biofuel feedstock may also be transported considerable distances to conversion facilities after harvest, further increasing the potential for release of propagules into natural areas (Barney and DiTomaso 2010).

Given the projected widespread planting of biofuel crops, distinguishing between species or taxa that are likely to become invasive and those with low risk of invasion will minimize the potential for invasions and promote the sustainability of energy production from second-generation biofuel crops (Barney and DiTomaso 2008, Cousens 2008, Gordon *et al* 2011). Here we define an 'invasive species' as one that is both non-native to an ecosystem (i.e. would not have occurred without human action) and causes ecological or economic impacts (Lockwood *et al* 2007). We define 'invasion risk' as the likelihood that a species will become invasive. In this paper we review the current methods used to evaluate the invasion risk of non-native species, describe how biofuels provide a unique invasion risk situation, and provide specific recommendations for conducting experimental tests of invasion risk. Such tests have frequently been called for (e.g. Mack 1996, Barney and DiTomaso 2010) but thus far little guidance has been offered.

CURRENT APPROACHES FOR EVALUATING INVASION RISK

Biological invasions are the result of a non-native species progressing through a series of steps from introduction to naturalization to having widespread ecological impacts. As such, to become invasive a species must overcome the barriers between stages in the invasion process, and most approaches for predicting invasion risk focus on a species' ability to pass through specific barriers (figure 1). Tools to evaluate invasion risk range from qualitative tools such as expert evaluation to semi-quantitative tools, including weed risk assessments and climate matching, to quantitative, empirical tools that involve experimental tests in the field (figure 1).

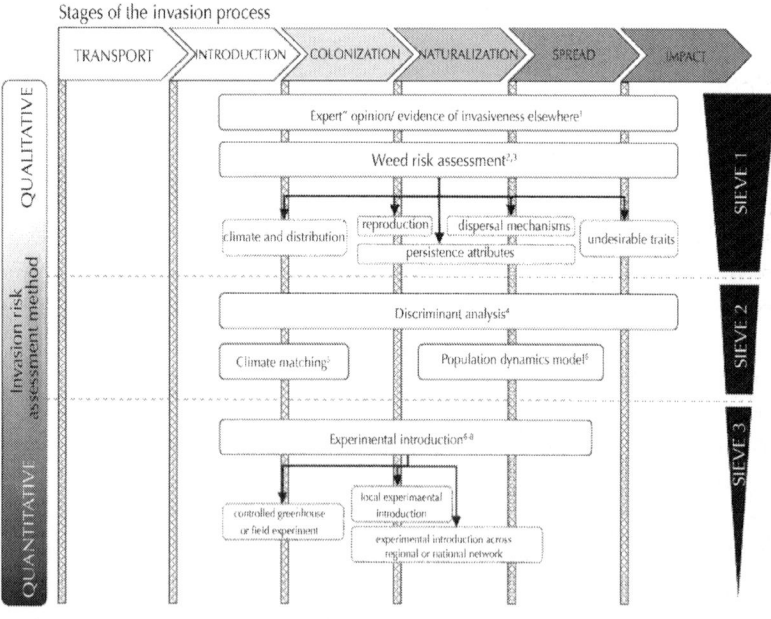

Figure 1: Tools to evaluate the invasion risk of biofuel crops presented on a gradient from qualitative to quantitative methods across the stages (shaded arrow boxes) and barriers (hatched bars) of the invasion process. Tools near the qualitative end of the spectrum are characterized by dependence on lit-

erature-derived information while quantitative tools are focused on empirical data, experimental tests, and statistical analyses. The gradient also corresponds to the three-sieve approach to invasion risk assessment described by Davis et al (2010). Notes: [1]FAO (2004), [2]Gordon et al (2011), [3]Daehler et al (2004),[4]Richardson and Rejmánek (2004), [5]Rodda et al (2007), [6]Davis et al (2011), [7]Minton and Mack (2010), [8]da Silva et al (2011).

The most basic of the existing invasion risk assessment tools is qualitative expert assessment, where experienced individuals use their knowledge to assess a species' likely invasiveness (Hulme 2011). The simplest of these approaches relies on the question of whether the species is invasive elsewhere outside its native range. In support of this approach, history of invasiveness correctly predicts invasion probability in a new range 90% of the time (Panetta 1993, Gordon et al 2008b). However, the rapid increase in novel taxa (e.g. for biofuel crops) through hybridization, traditional crop breeding, and genetic modifications, means that the novel taxon may not have a sufficiently long introduction history elsewhere that can be used for predictive purposes. Additionally, the novel taxa may demonstrate such fundamental differences from the original or parent taxon that assessment of the original taxon is of limited value.

One of the most widely used risk evaluation tools is the Australian Weed Risk Assessment (WRA). The WRA is a semi-quantitative tool such that points associated with positive or negative answers to 49 questions about a species' current weed status in other parts of the world, climate and environmental preferences, and biological attributes are summed, and the total score is used to assign a high or low risk of invasion or the need for further information (Pheloung et al 1999). The WRA is considered a semi-quantitative tool because there are scores associated with predictions that allow users to identify the probability of error associated with the different outcomes (Gordon et al 2012). Variations of the WRA have been successfully adapted for several climatic regions (e.g. Gordon et al 2008a, Buddenhagen et al 2009). Across geographies, the WRA can correctly identify major invaders and non-invaders 90% and 70% of the time, respectively (Gordon et al 2008a), and it can be performed quickly with relatively little expertise. Results are also relatively consistent across geographies of similar climate (Chong et al 2011).

Despite the utility and accuracy of the WRA, there are limitations. Where new taxa are involved, one has to assume that the only traits that might differ from the parent plant(s) or wild type are those that were the subject of manipulation (e.g. sterility, cold tolerance, etc). However, stochastic processes and local interactions with biotic and abiotic features of ecosystems in a new range can greatly influence the establishment and performance of taxa (Hulme 2011, Minton and Mack 2010). These complex processes and relationships are often nested, non-linear, and affected by temporal lags and positive feedbacks (Hulme 2011). Thus they are difficult to predict from the information used in the WRA. Additionally, the WRA does not always produce a conclusion about invasion risk. Even after implementation of a secondary screen (Daehler *et al* 2004) for species requiring further evaluation, on average 10% of species remain in that category (Gordon *et al* 2008a). Furthermore, concerns have arisen about WRA assessor subjectivity and inconsistency (Davis *et al* 2011, Hulme 2011). Finally, because there is no penalty for answering 'unknown' for questions that lack supporting evidence, and only ten questions need to be answered to draw a conclusion, it is possible for the WRA to incorrectly indicate low invasion risk for species whose basic biology is relatively undocumented (Barney and DiTomaso 2008).

There may be specific limitations of the WRA for evaluating the invasion risk of biofuel crop species (Cousens 2008). The interpretation of commonly used terms 'cultivation' or 'domestication' can become unclear in the context of biomass crops. For example, Question 1.01 in the assessment asks, 'Is the species highly domesticated?'. Assessors are instructed to answer 'yes' if the taxon has been intentionally selected over several generations for a particular trait or suite of traits that likely reduces weediness. This decreases the final score by three points, making it more likely that the plant will be accepted for introduction. However, answering 'no' does not penalize the final score towards a 'reject' outcome. Crops that are cultivated for biofuel feedstock differ from annual species that are primarily cultivated for food production in that breeding for ideal biomass crops may increase typical invasive characteristics (Barney and DiTomaso 2008). This difference may result in underestimation of invasion risk for proposed biofuel crops that are close to the score cutoffs of invasion risk categories.

Weed risk assessment is sometimes conducted using other quantitative prediction tools that make use of demographic parameters derived from the literature or field studies in order to produce population dynamics models. In such demographic population models, invasion risk is determined by the population's ability to reproduce faster than the mortality rate (i.e. net population growth rate, λ). If these models output a $\lambda > 1$, this would indicate the potential for naturalized populations to increase, and to potentially become invasive (figure 1, Davis et al 2011). However, such models are only applicable to the conditions and habitats evaluated in a particular study. Similarly, the application of discriminant functions derived from life history and biological characteristics (e.g. height, per cent germination, and minimum juvenile period) can be used to calculate a 'Z-score', which can be used to separate invaders from non-invaders based on their probability of becoming widespread (figure 1, Richardson and Rejmánek2004). Invasion risk can also be assessed using climate data. Climate matching uses algorithms to identify locations that are at risk of being colonized by a non-native species on the basis of similarity to the temperature and rainfall parameters of the species' native range (Rodda et al 2007). While climate characteristics and invasive potential do not always have a direct relationship, this tool gives a good indication of the potential for non-native species to colonize a new range (figure 1, Barney and DiTomaso 2011). Finally, a tool was recently developed to evaluate invasion risk based on biological traits but scaled to consider economic benefits (Schmidt et al 2012). These approaches are quite valuable; however, the same difficulties arise with these tools as with the WRA when few life-history details for a species or cultivar are known.

These limitations suggest the need for a more holistic approach to screening taxa of potential economic value, like biofuels, where a fully precautionary approach may be overly restrictive. To that end, Davis et al (2010) proposed that new biofuels be subjected to screening with multiple sieves that might include all of the previously discussed tools, beginning with the WRA. If the WRA (sieve 1) is unable to produce an 'accept' or 'reject' outcome, quantitative predictive tools such as population dynamics models and climate matching (sieve 2) should be used. If the first two sieves conclude that a species is conditionally acceptable then they recommend that the proposed biofuel species should be further evaluated using experiments (sieve 3). Others have suggested that risk assessments should not be used for final conclusions on invasion risk and that crops should be further screened using in situagronomic trials (Barney and DiTomaso 2008, 2010). Further, it has been suggested that assessment be performed for all candidate genotypes and cultivars because ecological interactions and growth characteristics (i.e. invasion potential) can vary widely within a species (Casler *et al* 2004). Here we outline the advantages, limitations, and potential research methods for such experimental tests of invasion risk.

EXPERIMENTAL TESTS OF INVASION RISK

Experiments that evaluate the invasion risk of biofuel crop species and taxa will increase confidence in predictions of that risk by providing an additional source of information for tools such as the WRA and may resolve recommendations for species that fall into the WRA 'evaluate further' category. Several previous studies have experimentally assessed factors contributing to invasion risk. For example, Myers (1983) used a combination of greenhouse experiments and seeding trials in different habitats to predict sites that were susceptible to invasion by the non-native tree *Melaleuca quinquenervia* in southern Florida. More recently, Minton and Mack (2010) conducted field trials to investigate how founder population size and the amount of cultivation influenced the invasion potential of four non-native species. Davis *et al* (2011) expanded upon these approaches in an experiment to determine the role of disturbance and timing of introduction for the invasion success of the biofuel candidate *Camelina sativa* in rangeland ecosystems. While the value of experimental work is exemplified in these studies, a general framework for evaluating the invasion risk of biofuel crops is needed.

We draw upon the common threads of past experimental evaluations of invasion potential and make recommendations for future work. Because current invasion risk assessment methods poorly evaluate the ability of a species to spread locally and generate self-sustaining populations (figure 1), we focus on this barrier to invasion. Our recommendations include a series of incremental steps from small-scale experiments to determine a species' basic biological requirements, to experimental introductions across broad geographic networks (figure 2). This stepwise approach should increase confidence in the evaluation of invasion risk of introduced biofuel crop species or taxon (hereafter 'species').

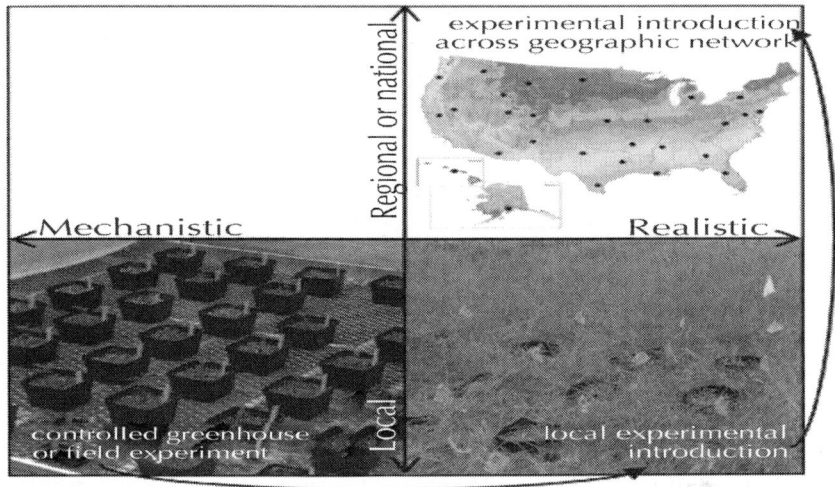

Figure 2: Recommended stages of experimental evaluation of potential biofuel crops on a gradient from mechanistic to realistic and from local to regional or national experiments. Arrows follow the suggested progression from local mechanistic experiments to trials in natural areas and eventually experimental introductions across broad geographic networks. The map shows USDA plant hardiness zones (http://planthardiness.ars.usda.gov) and points indicate Plant Materials Centers and Service Areas (http://pplant-materials.nrcs.usda.gov/centers).

Greenhouse, Growth Chamber, and Local Field Trials

The goal of small-scale, local experiments is to identify conditions that are favorable for germination, survival, and growth of proposed biofuel species. While experiments in such controlled environments cannot fully represent the more complicated invasion process, they can be valuable for informing subsequent experiments, which aim to identify habitats most susceptible to invasion. Further, when little is known about the basic biology and traits related to invasiveness of a species, contained experiments are an appropriate precautionary approach prior to field introductions. Preliminary common garden trials in a controlled outdoor setting may also be useful for determining requirements for establishment such as frost protection and insecticide application, which are difficult to approximate in a greenhouse (figure 2). Experimental treatments in growth chambers or greenhouses might include varying levels of moisture, nutrients, light, soil types, or herbivores and pathogens that are hypothesized to be important for determining establishment success.

Small-scale experiments have been used previously to better understand invasion risk of non-native species. For example, Myers (1983) evaluated how different water regimes affected time to germination, seedling height and biomass of *M. quinquenervia*. Treatments in their greenhouse experiment included seven different combinations of watering amount and frequency, and drainage frequency that corresponded to natural hydroperiod regimes found in different habitats in nature. They found that seedlings could survive extended periods under water and thus determined that areas with periodic flooding should be included as sites for their experimental field introductions (Myers 1983). This work offers a useful model for initial experiments, but the approach taken for evaluating non-native biofuel candidates will need to be tailored to the species or plant functional group (e.g. grass, shrub, or tree) of interest. For example, germination and establishment requirements often vary widely among small-seeded grasses and larger-seeded trees, which differ greatly in stored resources that can be important for overcoming soil and litter barriers or immersion. In addition, propagule type (stem cuttings/fragments versus seed) may greatly influence when and where individuals might establish.

In general, the detail and extent of small-scale trials will vary among species and functional groups depending on how much is known about their biology. In addition, if multiple specific cultivars are proposed for introduction, each cultivar will need to be evaluated individually. General information about a species may accelerate this process by providing information about what factors are most important for establishment success, but we urge caution in applying results from one cultivar or genotype to other taxa. More broadly, information generated in small-scale experiments is vital for identifying habitats that are most susceptible to invasion, thereby increasing the efficiency of experimental introductions.

Experimental Introductions

The next step in experimental evaluation is to assess the most important factors for establishment, survival, and reproduction in experiments that most closely simulate natural invasions (figure 2). Propagules including seeds, or stem fragments if the species reproduces vegetatively, should be scattered into test plots on the soil surface to mimic natural dispersal by wind or water, or disturbance treatments might be applied that result in propagules being buried by soil or leaf litter. We recommend collecting various performance measures include seedling germination and emergence, survival from emergence to maturity, and seed production. These measures can be used to calculate population growth rate (λ), which can be compared among species (Parker and Kareiva 1996), and provide a measure of a species› ability to overcome reproductive limitations associated with barriers to naturalization and spread (figure 1). Evaluation of individual life-history stages using sensitivity analysis (Koop and Horvitz 2006) might help to determine best management practices (BMPs) to control or prevent naturalization and spread. However, it may not be possible to measure fecundity in limited experimental time frames for certain species, particularly for shrubs and trees. More importantly, allowing a potential invasive species to flower and reproduce may not be advisable depending on the location and extent of experiments. Even if fecundity measures cannot be attained, measurement of response variables should be continued for a minimum of two years, although we suggest collecting up to four years of data to overcome the influence of stochastic weather conditions.

Evaluation of invasion risk needs to be based on both the attributes of introduced species and the specific environmental context; thus controlled introductions should occur at multiple sites that represent the range of land use types and natural communities that might be subjected to introductions (figure 2; Ewel *et al* 1999, Parker and Kareiva 1996). Previous experimental work has demonstrated the value of repeating experiments in different environments. For example, Davis *et al* (2011) employed two historically disturbed rangeland ecosystems that differed in soil characteristics and species richness and observed that the emergence of the introduced non-native species differed between sites. Likewise, *M. quinquenervia* introduced into various community types (e.g. burned transition, drained pond cypress, bald cypress-mixed hardwood, and mangrove) only survived for three years in certain habitats (Myers 1983). Thus, while limited introductions into a few sites may provide some insights, these results suggest that the full range of potentially susceptible habitats needs to be evaluated in order to determine when and where invasions might occur.

Experiments might first be located at only one location or at several locations spaced at latitudinal intervals, but would eventually need to occur in potentially susceptible environments across the full regional or national target range for biofuel planting and dispersal potential (figure 2). Testing across a broad geographic network is valuable not only because invasion potential would be determined relative to different and changing climates, but also because it can increase predictive power in a short time frame by increasing spatial diversity. One possibility is that US Department of Agriculture (USDA) Plant Materials Centers and Service Areas (figure 2, http://plant-materials.nrcs.usda. gov/centers) could provide the distributed field sites appropriate for these tests. Information gained from experiments conducted across the regional or national network could then be integrated in a Geographic Information System (GIS) framework to provide predictive maps of invasion risk (Lindgren 2012).

In addition to testing in different community and land use types, we also recommend that experiments consider the potentially important factors of disturbance, founder population size, and timing of introduction, by comparing these factors as treatments in split plot or block designs (Minton and Mack 2010, Davis et al 2011, da Silva et al 2011). Disturbances such as changes in grazing pressure, fire regime, soil scouring, or removal of vegetation are known to be important drivers of invasion (Elton 1958). Further, the type of disturbance and the habitat in which they occur (Burke and Grime 1996) can both affect invasion potential. In an experimental introduction, a disturbance treatment could be applied to individual plots (Davis et al 2011) or by introducing seeds into plots at sites where the entire area has been disturbed naturally (Myers 1983). Using natural disturbances can accurately represent processes that occur in nature but it may prove difficult to find replicates and appropriate controls across a broad region.

Propagule pressure is also recognized as an important factor in facilitating invasion (Lockwood et al 2005). Experimental introductions should test different size founder populations in order to predict the level of propagule pressure needed to overcome stochastic processes and local circumstances that might affect establishment success (Minton and Mack 2010). The number of individuals introduced should reflect hypothesized maximum and intermediate levels that could occur naturally. The timing of introduction should also correspond to the periods of time when propagules are most likely to be spread, such as during seed production and harvest and planting events. Ideal experimental introductions to assess invasion risk might occur in both spring and fall (Davis *et al* 2011, Parker and Kareiva 1996) using seed and potentially other propagules, and would be repeatedly tested over multiple years.

The development of BMPs that reduce invasion risk, which could be voluntary or regulated, are another avenue for investigation that can be coupled with introduction trials. Some practices may be advisable regardless of the species involved. For example, creating a cleared buffer around cultivation sites would allow for initial monitoring of spread. Such cleared areas could be surrounded by similar buffer zones with dense vegetation (e.g. turfgrass). The width of these buffers would be dependent on the likely distance of vegetative growth or seed dispersal, but might be 10 m for herbaceous species and 30 m for woody species (Cremer 1977). Additional practices to reduce invasion risk would be dependent on more specific establishment traits of the species involved. One example of a BMP designed to reduce the probability of invasion has been developed by the University of Florida for five cultivars of *Eucalyptus grandis*. This BMP stipulates: (1) 75 ft planting distance from waterways, ditches, and wetlands; (2) planting in monoclonal blocks; (3) maintaining densely vegetated or bare soil perimeter buffers of 75 ft; (4) harvesting prior to seed maturation; and (5) monitoring for and eliminating seedlings within 200 ft of cultivation site (http://plants.ifas.ufl.edu/assessment/conclusions). Development of BMPs for other species and taxa could be accomplished by modifying this model using information derived from local trials and experimental introductions.

LIMITATIONS OF EXPERIMENTS

Empirical methods will greatly increase our ability to assess the invasion potential of non-native bioenergy crops but there are limitations. Compared to non-empirical methods for assessing invasion risk, experiments require greater investments in time, money, labor, and space. As such, it has been suggested that experimental screening is not practical for all non-native or genetically engineered organisms proposed for introduction (Parker and Kareiva 1996). However, we believe that it is reasonable to suggest that experimental screening should be performed for the entire, relatively small, pool of proposed non-native biofuel crop taxa. Questions still remain about who will be responsible for the cost and oversight of experimental evaluations, but one scenario is that evaluations are required as part of a standardized permitting process at the state or preferably federal level. Alternatively, such testing could be promoted as a voluntary code of conduct similar to compliance to best management practices for pesticide and fertilizer use in forestry (Chimera et al 2010). The expense of conducting evaluations could be the responsibility of the importer or developer of the biofuel feedstock or others in the industry that will profit from the introduction. This approach is analogous to the testing of genetically modified organisms in the United States, which are regulated under the National Environmental Policy Act (Parker and Kareiva 1996).

One of the other primary concerns of experimental evaluation is that introductions might result in invasions of non-native species into natural areas. However, in locations with particularly sensitive natural areas, risk of escape can be reduced by placement of experiments within an agricultural or developed landscape distant from conservation areas and by adopting cautionary measures to prevent individuals from spreading outside of the experimental introduction sites. Such actions should include constructing silt fencing or other barriers around plots to reduce the movement of seed, vegetation buffers, monitoring surrounding areas for escaped individuals, and removal of all individuals at the conclusion of the experiment. Experimental work with species with long-lived seedbanks or long distance propagule dispersal mechanisms would require additional precautions. Regardless of the design, all experiments should include protocols for monitoring and responding to off-site colonization—both for assessment and control of incipient spread.

Finally, caution is required when drawing conclusions about invasion risk. It is not logistically possible to test all of the possible scenarios under which establishment and invasion might occur and inferences can only be made confidently for locations with similar biotic and abiotic features to the experimental sites. Additionally, experiments may not be able to capture relatively rare events that might influence invasion risk such as weather extremes or pest outbreaks. Researchers will need to balance what is possible to test with the certainty of their conclusions, keeping in mind that our intent is to improve the probability with which we assess risk, rather than identify definitive outcomes.

CONCLUSIONS

Current risk assessment systems offer a useful starting point for evaluating the invasion risk of biofuel crops but are limited in their utility due to unavoidable uncertainty. Thus, more experimental, repeatable, geographically distributed tests will improve our ability to evaluate invasion potential (Davis *et al* 2010). While we will not be able to guarantee that 'acceptable' biofuel crops will never become invasive, experimental screening increases our ability to predict invasiveness. If the risk is high, the challenge will be to assess whether the cost of cultivation of a species is likely to exceed the potential benefit. This level of risk assessment will require coordination among all stakeholders involved, including agronomists, ecologists, industry experts, and economists. Information on invasion risk will have to be integrated into a larger regulatory framework at the state or national level in order to maximize economic benefits but minimize the risk of ecological damage. Through implementing experimental screening of invasion risk for non-native species or cultivars of biofuel crops, we can stimulate progress in the sustainable development of second-generation biofuels, and simultaneously promote practices that protect natural areas from potential invasions.

ACKNOWLEDGMENTS

Publication of this letter was funded in part by the University of Florida Open-Access Publishing Fund.

REFERENCES

1. Barney J N and DiTomaso J M 2008 Nonnative species and bioenergy: are we cultivating the next invader? *Bioscience* 58 1–7

2. Barney J N and DiTomaso J M 2010 Invasive species biology, ecology, management and risk assessment: evaluating and mitigating the invasion risk of biofuel crops *Plant Biotechnology for Sustainable Production of Energy and Co-Products* ed P N Mascia, J Scheffran and J M Widholm (Berlin: Springer)

3. Barney J N and DiTomaso J M 2011 Global climate niche estimates for bioenergy crops and invasive species of agronomic origin: potential problems and opportunities *PLoS One* 6 e17222

4. Buddenhagen C E, Chimera C and Clifford P 2009 Assessing biofuel crop invasiveness: a case study *PLoS One* 4e5261

5. Burke M J and Grime J P 1996 An experimental study of plant community invasibility *Ecology* 77 776–90

6. Casler M D, Vogel K P, Taliaferro C and Wynia R L 2004 Latitudinal adaptation of switchgrass populations *Crop Sci.* 44293–303

7. Chimera C G, Buddenhagen C E and Clifford P M 2010 Biofuels: the risks and dangers of introducing invasive species*Biofuels* 1 785–96

8. Chong K Y, Corlett R T, Yeo D C J and Tan H T W 2011 Towards a global database of weed risk assessments: a test of transferability for the tropics *Biol. Invasions* 13 1571–7

9. Cousens R 2008 Risk assessment of potential biofuel species: an application for trait-based models for predicting weediness *Weed Sci.* 56 873–82

10. Cremer K W 1977 Distance of seed dispersal in eucalypts estimated from seed weights *Aust. For. Res.* 7 225–8

11. Daehler C C, Denslow J S, Ansari S and Kuo H C 2004 A risk-assessment system for screening out invasive pest plants from Hawaii and other Pacific islands *Conserv. Biol.* 18 360–8

12. Dale V H, Kline K L, Wright L L, Perlack R D, Downing M and Graham R L 2011 Interactions among bioenergy feedstock choices, landscape dynamics, and land use *Ecol. Appl.* 21 1039–54

13. da Silva P H, Poggianai F, Sebenn A M and Mori E S 2011 Can Eucalyptus invade native forest fragments close to commercial stands? *For. Ecol. Manag.* 261 2075–80

14. Davis A S, Cousens R D, Hill J H, Mack R N, Simberloff D and Raghu S 2010 Screening bioenergy feedstock crops to mitigate invasion risk *Front. Ecol. Environ.* 8 533–9

15. Davis P B, Menalled F D, Peterson R K D and Maxwell B D 2011 Refinement of weed risk assessments for biofuels using*Camelina sativa* as a model species *J. Appl. Ecol.* 48 989–97

16. Demirbas A 2009 Political, economic, and environmental impacts of biofuels: a review *Appl. Energy* 86 S108–17

17. Elton C S 1958 *The Ecology of Invasions by Plants and Animals* (London: Methuen)

18. Ewel J J *et al* 1999 Deliberate introductions of species: research needs *Bioscience* 49 619–30

19. FAO (Food and Agrigulture Organization of the United Nations) 2004 *International Standards for Phytosanitary Measures Publication No. 11. Rev. 1; Pest Risk Analysis for Quarantine Pests Including Analysis of Environmental Risks*

20. Gordon D R, Gantz C A, Jerde C L, Chadderton W L, Keller R P and Champion P D 2012 Weed risk assessment for aquatic plants: modification of a New Zealand system for the United States *PLoS One* 7 e40031

21. Gordon D R, Onderdonk D A, Fox A M and Stocker R K 2008a Accuracy of the Australian weed risk assessment system across varied geographies *Diversity Distrib.* 14 234–42

22. Gordon D R, Onderdonk D A, Fox A M, Stocker R K and Gantz C A 2008b Predicting invasive plants in Florida using the Australian Weed Risk Assessment *Invasive Plant Sci. Manag.* 1 178–95

23. Gordon D R, Tancig K J, Onderdonk D A and Gantz C A 2011 Assessing the invasive potential of biofuel species proposed for Florida and the United States using the Australian Weed Risk Assessment *Biomass Bioenergy* 35 74–9

24. Hulme P E 2011 Weed risk assessment: a way forward or a waste of time? *J. Appl. Ecol.* 49 10–9

25. Koop A L and Horvitz C C 2006 Population dynamics and invasion rate of an invasive tropical understory shrub *8th Int. Conf. on the Ecology and Management of Alien Plant Invasions* (Raleigh: EMAPi)

26. Lindgren C J 2012 Biosecurity policy and the use of geospatial predictive tools to address invasive plants: updating the risk analysis toolbox *Risk Anal.* 32 9–1

27. Lockwood J L, Cassey P and Blackburn T 2005 The role of propagule pressure in explaining species invasions *Trends Ecol. Evol.* 20 223–8

28. Lockwood J L, Hoopes M F and Marchetti M P 2007 *Invasion Ecology* (Malden: Blackwell)

29. Mack R N 1996 Predicting the identity and fate of plant invaders: emergent and emerging approaches *Biol. Conserv.* 78 107–21

30. Mack R N 2008 Evaluating the credits and debits of a proposed biofuel species: giant reed (*Arundo donax*) *Weed Sci.* 56 883–8

31. Minton M S and Mack R N 2010 Naturalization of plant populations: the role of cultivation and population size and density *Oecologia* 164 399–409

32. Myers R L 1983 Site susceptibility to invasion by the exotic tree *Meleleuca quinquenervia* in southern Florida *J. Appl. Ecol.* 20 645–58

33. NISC (National Invasive Species Council) 2009 *Paper 11; Biofuels: Cultivating Energy, Not Invasive Species*(Washington, DC: NISC)

34. Panetta F D 1993 A system for assessing proposed plant introductions for weed potential *Plant Prot. Q.* 8 10–4

35. Parker I M and Kareiva P 1996 Assessing the risks of invasion for genetically engineered plants: acceptable evidence and reasonable doubt *Biol. Conserv.* 78 198–208

36. Pheloung P C, Williams P A and Halloy S R 1999 A weed risk assessment model for use as a biosecurity tool evaluating plant introductions *J. Environ. Manage.* 57 239–51

37. Raghu S, Anderson R C, Daehler C C, Davis A S, Wiedenmann R N, Simberloff D and Mack R N 2006 Adding biofuels to the invasive species fire? *Science* 313 1742

38. Richardson D M and Blanchard R 2011 Learning from our mistakes: minimizing problems with invasive biofuel plants*Curr. Opin. Environ. Sustain.* 3 36–42

39. Richardson D M and Rejmánek M 2004 What attributes make some plant species more invasive? *Ecology* 77 1655–61

40. Rodda G H, Reed R N and Jarnevich C S 2007 Climate matching as a tool for predicting potential North American spread of brown tree snakes *Proc. Managing Vertebrate Invasive Species* ed G Witmer and K Fagerstone (Fort Collins, CO: National Wildlife Research Center)

41. Ross M 2011 New *Miscanthus* development possible biomass game-changer? *FarmWeek 16 May* p 9

42. Schmidt J P, Springborn M and Drake J M 2012 Bioeconomic forecasting of invasive species by ecological syndrome*Ecosphere* 3 46

Organic Agriculture, Sustainability and Consumer Preferences

Errence Thomas[1] and Cihat Gunden[2]

T[1]North Carolina Agricultural and Technical State University, Department of Agribusiness, Applied Economics and Agriscience Education, USA

[2]Ege University, Faculty of Agriculture, Department of Agricultural Economics, Izmir, Turkey

INTRODUCTION

Scholars acknowledge that early man provided food for himself and his family via gathering what was available to him in his surroundings; he relied on nature for his sustenance. As hunter gatherers, man lacked the capacity to manipulate the environment to produce food beyond the amount that was available naturally. Consequently, there was minimal or no environmental impact, the human population remained small and in balance with nature; hunter gatherers' population could not expand beyond the available sources of food [1-3]. Over time, however, as hunter gatherers learn to cope with their environment and became more adept at gathering food, the population increased, leading to the next stage in the evolution of the food production system—the Neolithic revolution or the development of agriculture. The development of agriculture led to sedentary communities, increase in population size and the specialization of labor, all of which facilitated technological development, i.e., improved tools, dwellings and means for transporting water and materials. In sum, man learned and applied techniques for domesticating animals and plants, or put another way, agriculture was invented. Yet, at this early stage in the practice of agriculture, man's interaction with his sustenance base could be described as "give and take"; a relationship in which man essentially learned from his experience living in the environment, a sort of 'symbiotic" relationship with his sustenance base that resulted in little or no adverse environmental impact. Even when there was adverse impact, the population was small and technology environmentally benign, which allowed the sustenance base to recover. The invention of agriculture laid the foundation for the development of civilization, increase in knowledge and man's capability to manipulate the environment. It was not until the birth of modern science and its application to the development of techniques for producing goods and services that man acquired the capability to manipulate the environment for producing food to meet his needs. The birth of modern science, following the Enlightenment, nurtured a culture that promoted and reinforced the world view that man through the application of science would be able to master and manipulate the environment to meet his needs. Advances in science during this era (17th and 18th century) led to the Industrial Revolution and the progressive industrialization of agriculture.

Prior to the intensive application of science to agriculture, the production of food and fiber relied on what is now referred to as traditional methods, which included: crop rotation, organic manure from animals and cover crops, animal power, intensive use of labor on small farms and a conventional artisan approach to plant and animal improvement—agriculture relied heavily on natural process, i.e., the ecology in which it was nested. Thus, in terms of today's language food production was substantively organic. The industrial revolution transformed traditional agriculture with: (1) the application of farm machinery for land preparation, reaping, hauling, irrigating, land clearing, fertilizer, manure and pesticide application; (2) the development and application of fertilizers, insecticides and weedicides; (3) application of sophisticated irrigation systems; (4) the application of principles of genetics to plant and animal breeding and (5) the practice of monoculture. These technologies have led to staggering increases in crop and animal production and productivity, larger farms and fewer farms and farmers [1-2, 6] and increased negative impact on the sustenance base [1-2,6-8]. Another phase of agricultural evolution involved the application of information technologies, biotechnologies and modern science-based business management practices to organize and operate food production systems, leading to further gains in efficiency and productivity. Striking features of this phase include the following: large corporate style farms, drastic decline in family farms and profound innovations in the application of biotechnologies to the improvement of plants and animals. The progressive evolution of man's food gathering and food production relationship with his sustenance base (the ecology or environment) is characterized by: (1) his increasing capacity to apply science in developing the technologies used to manipulate the sustenance base or the ecological capital to meet his needs for food and fiber; and (2) the progressive ecological impact of these technologies. Prior to the phase of intensive application of science to agriculture, food production could be described as nature-based with food production and population more or less in balance with nature.

THE IMPACT OF AGRICULTURE ON THE ENVIRONMENT

Rachel Carson's seminal work "Silent Spring" documented the environmental impact of insecticide on the environment [9]. Other authors including [1-2, 6-8] have documented an increasing environmental impact of conventional industrial agricultural technologies. Among the major impacts are point and non-point pollution from fertilizers and pesticides use; deforestation; desertification; salinization; soil erosion and sediment deposition downstream; degradation of water aquifers, accumulation of toxic compounds, loss of biodiversity; and habitat fragmentation. The net effect of these impacts over time will be to reduce the capacity of the sustenance base to support increases in food production to meet the needs of future generations and the needs of those who currently suffer from hunger and malnutrition.

These concerns regarding health, as well as the environmental impacts and sustainability of conventional industrial agriculture have led to efforts directed at developing more sustainable alternatives as described by [10-13]. Alternatives, variously described as organic food production systems, community supported agriculture (CSA), community-based agriculture, and civic agriculture have begun to resonate and garner significant public support. These alternative approaches to food production are community-based food production systems. Community-based agriculture initiatives are nature-based and produce food in an environmentally sustainable manner [14-15]. Sustainable agricultural production systems practice crop rotation, no-till farming, diverse cropping patterns, use of organic matter or organically derived fertilizers, integrated pest management, biological control, cover cropping, timing of planting, leaving land in fallow, a variety of water conservation techniques and make optimum use of the natural biological cycles. The objective of a sustainable agricultural system is to forge a symbiotic relationship with the ecological capital and in the process learn to use the resources it provides without affecting the capacity of the ecological capital to support food production. This approach is tantamount to using a portion of the interest from an investment portfolio and ploughing back some earnings to ensure the continued productive capacity of the base investment capital. In contrast, conventional industrial agriculture views the ecology as primary capital input or raw material that is to be manipulated or consumed in the production process. The focus of sustainability in food production is to develop a food production system that mirrors or integrates with the natural ecology in which it exists. It is believed that such a system would achieve the highest degree of sustainability--the capacity to persist through time as a system of food production.

SUSTAINABLE AGRICULTURE THE UNDERGIRDING PRINCIPLE OF ORGANIC AGRICULTURE

What exactly is sustainable agriculture? Scholars and technocrats alike don't agree on a single definition, primarily because: (1) there is no way a single definition of the concept could be applied to cover the diversity of ecologies, cultural and economic conditions under which agriculture is practiced, and (2) there are several stakeholders, with a vested interest in the concept, who cannot agree on a single definition [16]. Essentially then, the practice of sustainable agriculture will be defined by local ecological and social conditions. Even though there is lack of agreement on a single definition of sustainable agriculture, there is general agreement that conventional agriculture or industrial agriculture is not sustainable for reasons mentioned above. For example, conventional agriculture depends increasingly on energy supplies from nonrenewable sources, depends on a narrow genetic base and intensive use of chemical fertilizers and pesticides. In addition, it relies on subsidies and price support, has an increasing negative impact on the environment as evidenced by the loss of species, habitat destruction, soil depletion, consumption of fossil fuels and water-use at unsustainable rates, and contributes to air and water pollution and risks to human health [17].

Notwithstanding the difficulties involved in defining sustainable agriculture, given the threat posed by conventional agriculture, scholars still continue to work to define and clarify the concept. For example, Ikerd [18] proposed the following definition: "... capable of maintaining its productivity and usefulness to society over the long run...t must be environmentally-sound, resource conserving, economically viable and socially supportive, and commercially competitive" (p.30). In a later work Ikerd argued that sustainability should be thought of as a goal to be achieved rather than a static concept with a fixed definition. Even though Ikerd's view has considerable intuitive appeal, we believe that having a working definition clarifies what a concept represents and provides the information needed for identifying its constituent elements and distinguishing it from other concepts. Description of an object or thing provides insight into the nature of what that thing is and what it can do. Since what a thing can do depends on what it is, insights into its nature enables us to hypothesize about potential courses of action regarding that thing. Or, put another way, insights developed from clarifying the definition of a sustainable agricultural production system enables us to design courses of action to attain a sustainable food production system.

In this chapter, we draw on Ikerd's definition and the definition of sustainable development proposed by [19]. We define a sustainable agricultural production system as the practice of agriculture to produce food and fiber that meets the needs of the current population without compromising the capacity of the ecological capital, on which it depends, to support the needs of future populations. This means the nutritional, recreational and fiber needs of current populations must be met within the ecological limits of our natural resource base (ecological capital). The primary elements making up our definition are: (1) need, (2) time, (3) ecological capital, (4) equity, (5) population and (6) practice. From our perspective, the first element, "need" entails consuming resources to satisfy a physiological or physical requirement over time. Technically, a need is a necessity that is not satisfied in a single instance; it is a continuing requirement. In this sense, a sustainable agricultural system is one that is capable of persisting through time to meet current and future needs. The second element, "time" is a key concept, because sustaining anything means making sure that the particular thing persists through time. In the case of a sustainable agricultural system, it means managing our relationship with the ecological capital in such a manner that it will continue to meet our needs and the needs of future generations. The third element in our definition, "ecological capital," represents the resource base or the stock of natural assets that support life and food and fiber production. Our definition of ecological capital varies slightly from that offered by [1]. In our definition, we emphasize the biological base (the ecosystem) from which all natural services and goods are derived. Wright [1], on the other hand, defines it as the sum of goods and services provided by natural and managed ecosystems (agriculture) that are essential to human life and well-being. We chose to use the ecosystem or biological base because if the ecosystem is degraded or depreciated, its productive capacity and ability to support food production through a managed ecosystem (agriculture) will be much reduced.

The fourth element, equity, refers to the necessity to manage the endowment of ecological capital to meet the needs of the current generation without damaging its capacity to provide for future generations. In the context of our definition, the principle of equity also implies observing rules of fairness in the production, distribution and marketing of food and in exploiting other goods and services provided by our endowment of ecological capital. Population, the fifth element, refers to the current generation who consumes the goods and services produced from ecological capital, as well as future generations who will be consuming future products and services from the ecological capital. The attainment of a sustainable agricultural production system depends on the size of the population whose needs are to be met, the consumption level of the population, and the type of technology used in the production process. The final element, practice, deals with not only the technology employed in the production process but also the political, economic and social factors that impinge on and shape the sustainable agricultural production system. Given our definition, the question becomes: what insights for action can we draw? From our perspective, there are four primary insights (our illustrations below draw on the work of [1]): First, the population or people whose needs are to be met by a sustainable agricultural production system may be viewed from a dual perspective. People are the beneficiaries of a sustainable agricultural production system. Second, people are agents who must be proactive in defining what a sustainable food production system should be.

If a sustainable food production system is to be more than a theoretical abstraction, agents-the beneficiaries-must be able to operationalize the system to produce sustainable benefits. In operationalizing the concept of a sustainable agricultural production system, both values and knowledge play a central role in this process. Knowledge tells us about the ecosystem and how it supports agricultural production and what sort of sustainable development is possible, while our system of values guides us in making a choice once our options have been made clear. In this sense, moving from abstraction to implementation will be guided by the process illustrated in Figure 1 below. As illustrated in Figure 1, a sustainable food production system must be economically feasible "meaning such a system must be affordable and economically efficient. The sustainable food production system must also be socially desirable "indicating that it must be in sync with the cultural disposition and values of the agents or people it will serve. Consistent with this view, [17], reject approaches to sustainability that focus on the description and development of sustainable farming practices regardless of the socio-productive characteristics of the farming systems in which they are applied. Finally, a sustainable food production system must be in harmony with the ecology which supports it. If the food production system is discordant with, or in any way detrimental to the ecology that supports it, such a food system will not be sustainable.

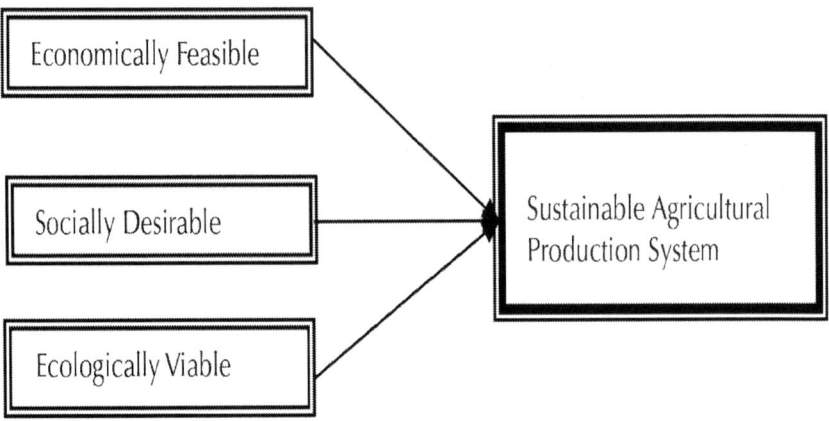

Figure 1: Sustainable food production system (adopted from [1]).

COMMUNITY AND SUSTAINABLE SYSTEMS

Third, to make a food production system sustainable following the precepts depicted in Figure 1, the agents of such a system must act according to the framework illustrated in Figure 2. This is the point where community plays a vital role in crafting and managing a food production system to achieve sustainable objectives.

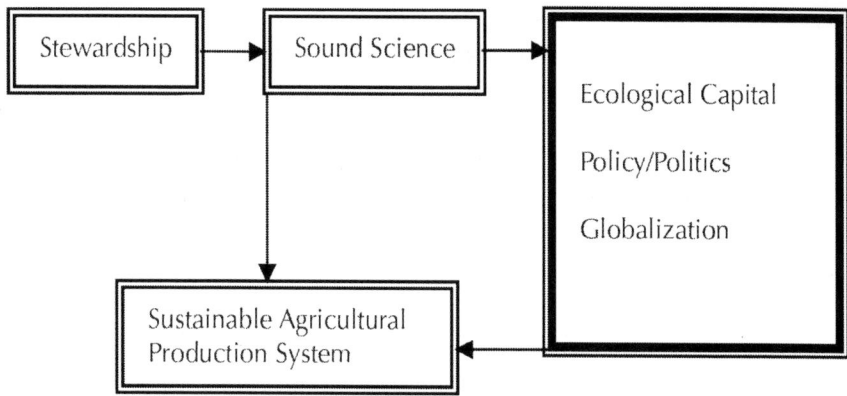

Figure 2: Framework for achieving a sustainable food production system (adopted from [1]).

In Figure 2, stewardship entails employing ethical principles and values in choosing how sustainability is achieved. For example, sound-science provides knowledge about the ecosystem and the possibilities for supporting agricultural pursuits in a sustainable manner. It also informs us about how to make good decisions through policies and the political process. Science generates knowledge about specific sustainable practices and their efficacy. It tells us about the impact of globalization on the distribution of food, trade, and the spread of pollutants and diseases. In sum, science tells us what is and what is not possible. Good stewards must apply ethical standards and values to choose from among the possibilities that science generates in designing and implementing a sustainable agricultural production system, and in evaluating and adjusting the system to meet sustainable objectives. So then, the pivotal question becomes: Who gets to choose from among the possibilities that science generates? Since food production in a sustainable system is inextricably linked to the local environment and the community's social and political infrastructure in which it exists, it follows that sustainable agricultural practices are defined by local ecological conditions and by the local social infrastructure which gives rise to the ethical values that guide stewardship. The connection of a sustainable food production system to ecological and social environments means that decisions concerning the design and development of sustainable agricultural production systems will have implications for everyone.

As a result, there will be several stakeholders with a vested interest in shaping the practice of sustainable agriculture. The reality is that citizens living in the same information rich environment as their leaders realize that the institutionalized bulwarks of authority are not omnipotent and that leaders are more or less ordinary people. Consequently, they assign less significance to the guidance of their leaders and institutions and have opted to become more reflective, proactive and self-regulating [20]. Implementing a sustainable agricultural production system in this context calls for collective action, because reflective and proactive citizens will insist on participating in the decision-making process. The support of diverse, reflective and proactive stakeholders is critical for ensuring that the values of stakeholders are reflected in defining and supporting the practice of sustainable agriculture.

Fourth, given that food systems depend on a healthy base of ecological capital regardless of their production technique, the sustainability of food systems can be conceptualized as existing on a continuum based on the level of integration with the natural ecosystem and the social environment in which it exists. At the high end of the continuum would be a production system that achieves the highest level of integration with the ecology and the social system in which it exists. And at the low end would be conventional/industrial agriculture. As indicated earlier, a sustainable system makes judicious use of available ecological capital by making optimal use of: biological cycles, the practice crop rotation, no-till farming, diverse cropping patterns, the use of organic matter or organically derived fertilizers, integrated pest management, biological control, cover cropping, timing of planting, leaving land in fallow and a variety of water and soil conservation techniques. To be sustainable, the food production system, as discussed earlier, must meet social and economic objectives within the limits of the ecology in which it exists. Sustainable food production must involve the community as consumers and stewards of the food production system. The system must also nurture and expand understanding of the interdependence of food production and the ecology which supports it. Considering that people are the agents and beneficiaries of a sustainable food system, communities must understand and accept that natural resources are finite, recognize the limits on economic growth, and encourage equity in resource allocation [17]. In other words, the drive for economic efficiency must be tempered by the need to preserve ecological capital and ensure social and economic equity. The trend toward large-scale Industrial profit driven farming has implications for the economic health of rural communities. For example, studies have demonstrated that independent hog farmers generate more jobs, more local retail spending, and more local per capita income than do larger corporate operations. Comparisons between conventional industrial agriculture and sustainable systems indicate that organic agriculture and sustainable systems are productive and economically competitive [17].

Given the concept of sustainable food system describe herein, we suggest that sustainable food systems exist on a continuum. The top end of the continuum would define a food production system that is nature-based and which achieves the highest level of integration with the ecology and social system in which it exits. We would label this highly ecologically and socially integrated food production system organic agriculture. Our conception of organic agriculture presented here is consistent with the definition proposed by Codex Alimentarius Commission which states that:

"Organic agriculture is a holistic production management system which promotes and enhances agro-ecosystems health including biodiversity, biological cycles, and soil biological activity. It emphasizes the use of management practices in preference to the use of off-farm inputs, taking into account that regional conditions require locally adapted systems. This is accomplished by using, where possible, cultural, biological, and mechanical methods, as opposed to using synthetic materials, to fulfill any specific function within the system." (Quoted in [21] pp.6)

At the low end of the continuum, displaying the lowest level of integration would be conventional industrial agriculture. Between these two extremes would be food production systems that manifest varying degrees of ecological and social integration or levels of sustainability. So then, organic agriculture is the ideal that we should work toward achieving as we strive to achieve a sustainable food system.

In today's market place there is a growing demand for organic products. And consumers seem willing to pay a premium price for products carrying organic quality labels. Questions that arise are how reliable are these quality labels and what level of confidence should consumers put in such labels? Usually the control process is carried out by independent certifiers who are guided by criteria promulgated by rule-making agencies. Certifiers must be vigilant and succeed in revealing departures from standards and opportunistic behavior in order for quality assurance labels to build up the reputation necessary to serve as a reliable quality signal. However, in the case of Potemkin attributes (where the desirable attribute is based on a process such as in organic production) there is the potential for quality statements to be made with little risk of disclosure of departures from standards, because consumer agencies, NGOs, and public authorities are usually not able to verify marketing claims or discover opportunistic behavior. What is needed to deter opportunistic behavior and identify departures from accepted standards is a quality monitoring protocol that covers the whole supply chain and ensures on-site inspections throughout the production process [22]. Another approach is to ensure stricter audit standards and rigorous training of certifiers, but these approaches are likely to increase the cost of certification and the resultant cost of organic products, which will drive down demand for products that are already offered to consumers at premium prices. In our view, a less expensive, organically-based and a more resilient approach would entail shortening the supply chain and fostering closer connection between producers and consumers.

We envisage that the community and farmers would fulfill the role of active co-stewards (the community of consumers and producers) of the organic food production system. As co-stewards of an organic food production system, farmers and consumers would be organized in networks that exchange ideas, share experiences and information and work together to solve problems. In this situation, an effective self-monitoring protocol that is grounded in a culture of trust and commitment to standards could emerge. The opportunity for farmers and consumers to interact as co-stewards would create an appreciation for the attributes that consumers' value, the relationship between these valued attributes, the production process and the price farmers are able to fetch for their product.

On the other hand, consumers would get an appreciation for the process that produces the valued attribute. Over time, the "deep trust" that would develop between producers and consumers as a result of the co-creation of understanding of the role of consumer and farmer in meeting each other's need would lead to an effective monitoring system. This level of understanding could potentially lead to the identification of points of weakness in the process; whereupon, co-stewards would take action to modify existing protocols that would reduce the likelihood of opportunistic behavior.

The idea of entrepreneurial social capital espoused by [23] provides a conceptual basis for our proposed co-creation of an effective and inexpensive monitoring system. In the instance outlined above, co-stewards (the community of consumers and producers) have the potential to serve as a catalyst for mobilizing entrepreneurial social infrastructure (ESI). [23] Defines ESI as having three elements: symbolic diversity, resource mobilization and quality of networks. Symbolic diversity enables co-stewards to encourage participation, dissent, accept challenges to the status quo and embrace constructive controversy and critiques; it encourages people to focus on the process and the arguments instead of the personalities involved. It also encourages resource mobilization, which involves promoting local investment by residents in the community, equity in resource and risk distribution and collective investment in the community. Quality networks are encouraged by establishing horizontal and vertical linkages. Horizontal networking links co-stewards in similar circumstances and promotes learning by sharing experiences and information from different perspectives. Vertical linkages draw on resources of others operating in dissimilar circumstances, or in different systems. *It enables co-stewards to attract resources from private and public sources outside the community, for example, from entities with different levels of expertise and capacity relevant to the problem at hand* (our emphasis).

ASSESSING CONSUMERS PREFERENCES TOWARD PRODUCTION SYSTEM AND CONSUMERS PREFERENCES FOR THE ATTRIBUTES OF FRESH FRUITS AND VEGETABLES

This next section will examine the attitude of consumers toward organic, sustainable and conventional production system and consumers preferences for the attributes of fresh fruits and vegetables. As discussed earlier, sustainable production lies between organic and conventional production system on our continuum described above. Thus, a sustainable agricultural production system is operationalized as employing good agricultural practices (judicious use of synthetic fertilizers and pesticides), integrated pest management and emphasizes the use of natural cultural practices and fertilizers and insecticides from natural sources as much as possible.

MEASURING PREFERENCES FOR FOOD PRODUCTION SYSTEMS

The advent of specialized stores offering organic produce and products and the allocation of supermarket self-space to organic produce and products attest to the increasing demand for food and food products produced under alternative production systems. The emergence of alternative food production systems and the discussion in the public domain concerning the health, environmental and social benefits they offer vis a vis conventional production systems may have, at the very least, sensitized consumers about the opportunities that exist for making food purchasing decisions based on the type of production system and its perceived benefits. Additionally, the promotion of healthy eating habits and the need for increased consumption of fruits and vegetables [26-28], plus the well-publicized need for environmental conservation [19] amplify the salience and relevance of differences between the food production systems in terms of their health, environmental and socio-economic impact. Consequently, our objective here is to assess consumer attitudes toward food produced under the following food production systems – conventional agriculture, sustainable alternatives and organic along five criteria – contribution to environmental conservation, food safety, food quality, contribution to wellness and contribution to community economic development by using Analytic Hierarchy Process (AHP).

Data and Methodology

The sample was designed following the protocol described by [29]. It was drawn proportionate to population size by county in Georgia,

North Carolina and South Carolina. After specifying the sampling frame parameter, the required sample was purchased from Survey Sampling Inc. Data were collected from a random sample of 252 respondents, which represents a cooperation rate of 30 percent. Researchers designed and formatted an analytic hierarchy questionnaire to collect data via a telephone survey. Enumerators asked consumers to compare three food production systems: conventional, sustainable and organic in terms of which consumers would prefer farmers to use in producing the fresh fruits and vegetables that they purchase or consume; taking into consideration environmental, food safety, food quality, wellness, and community development issues.

This study employed Analytic Hierarchy Process (AHP) to derive a measure of an individual consumer's preference for production systems in terms of the selected criteria which is consistent with previous research conducted in the U.S. [30]. The AHP, which was developed by [31], is one of the most commonly applied multi-criteria decision-making techniques. AHP is a subjective tool for analyzing qualitative criteria to generate priorities and preferences among decision alternatives (For more detailed information about AHP, see [32-34]. The AHP model, illustrated in Figure 3, was used to assess consumers' preferences for production systems in terms of environment, food safety, food quality, wellness and community development.

Cluster analysis was used to separate consumers into groups by: age, education and employment status. The aim of cluster analysis is to classify observations into relatively homogeneous groups called clusters, such that each cluster is as homogeneous as possible with respect to the clustering variables [35-36]. Researchers would then be able to determine if consumers' preferences for production systems varied by age, education or employment status. The Kolmogorov-Smirnov test was used to check whether the clustering variables were normally distributed and the Kruskal Wallis test was used to compare clusters.

Further analysis employing multidimensional scaling (MDS) was used to obtain "perceptual mapping of consumers' preferences for production systems. By transforming consumer judgments of overall preferences into distance represented in multidimensional space, MDS plots the three production systems and five criteria on a map such that those systems and criteria that are perceived to be very similar to each other are placed near each other on the map, and those systems and criteria that are perceived to be very different from each other are placed far away from each other on the map. In this way MDS provides a visual representation of the pattern of proximities (i.e., similarities) among the set of production system and the set of criteria employed in their assessment [36].

Results and Discussion

Consumers were grouped into three clusters. The mean of the variables used in the analysis is presented by the clusters in Table 1. There were statistically significant differences among clusters on the variables age, education and employment. The mean age (40.85) is the lowest in Cluster 1 and the highest (80.35) in Cluster 3. Education level is the highest (4.94) in Cluster 1 and lowest in cluster 3 (2.41). Employment status changes from employed in Cluster 1 (2.13) to unemployed in Cluster 3 (2.97). Cluster 1 is labeled "Young professional", while the cluster 2 and cluster 3 are labeled "Older-technician" and "Oldest-unemployed" respectively.

Table 1: Cluster analysis by age, education and employment

Variables	Clusters			Kruskal Wallis Test	
	1	2	3	Chi-Square	Asymp. Sig
Age	40.85	63.48	80.35	191.962	0.000
Education*	4.94	4.32	2.41	29.596	0.000
Employment**	2.13	2.52	2.97	69.077	0.000

[i] - * 1: Less than high school, 11: Professional/doctorate degree; ** 1: Part time, 2: Full time, 3: Unemployed

Table 2: displays the number of consumers by the clusters. The data show that 52.1 percent of consumers are "young professional", 33.5 percent are "older-technician", while the "oldest-unemployed" accounts for 14.4 percent.

Table 2: Consumer distribution by clusters

Clusters	Frequency	Percent	Cumulative Percent
Young professionals	123	52.1	52.1
Older technician	79	33.5	85.6
Oldest-unemployed	34	14.4	100.0
Total	236	100.0	

In the AHP Model, consumers were asked to assess conventional, sustainable and organic production systems, taking into account the ability of each to generate benefits related to environmental conservation, food safety, food quality, wellness and community economic development. The AHP model for assessing preferences for production systems in terms of these criteria is defined in Figure 3. The goal is to determine consumers' preferences for food produced under three production systems using the following criteria: environmental conservation, food safety, food quality, wellness and community economic development. These criteria are the perceived benefits generated by each system. In the AHP model illustrated below, consumers are being asked to choose their preferred food production system from among the alternatives: conventional, sustainable and organic production systems based on environmental conservation, food safety, food quality, wellness and community economic development criteria.

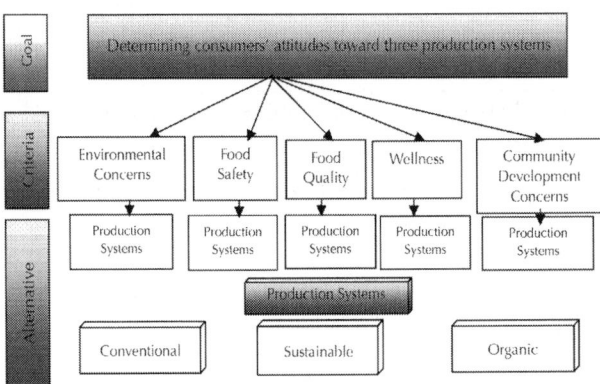

Figure 3: AHP model for consumer attitudes toward food production systems.

Table 3 shows the results obtained by applying the AHP model. The last column in Table 3 indicates consumers' average priority ratings for each criterion. The results indicate that consumers accorded priority in the following order to food safety (0.281) followed by wellness (0.275), food quality 0.209), environmental concerns (0.144) and community development concerns (0.091). Consumers considered food safety and wellness to be more important attributes or features of a food production system than other attributes such as food quality and the capacity of the food system to contribute to community development or environmental quality. In each row of Table 3, the preference scores for each type of production systems are presented. The third column of Table 3 shows that organic agriculture is preferred, when considered alone, based on its perceived capacity to generate benefits associated with wellness (0.575), food quality (0.533), safety (0.530), environmental concerns (0.515) and community development (0.514). The average preference rating of 0.544 shown in the last row of Table 3 indicates that consumers prefer the organic production system over the sustainable alternative and conventional agriculture, which were assigned preference ratings of 0.274 and 0.182 respectively.

Table 3: Consumers' attitudes toward food production systems by the criteria

Criteria	Conventional	Sustainable	Organic	Preference
Environmental Concerns	0.203	0.282	0.515	0.144
Food Safety	0.186	0.284	0.530	0.281
Food Quality	0.195	0.272	0.533	0.209
Wellness	0.162	0.262	0.575	0.275
Community Development Concerns	0.209	0.278	0.514	0.091
Final Decision	0.182	0.274	0.544	

[i] - [1] Consumer preference scores are ranged between 0 and 1. The sum of each row, excluding the preference in the last column, is equal to 1.00.

Since consumers' preferences for the production systems of food may be influenced by their demographic traits and behaviors [37], demographic traits may be used, where heterogeneity in consumers preferences exists, to segment consumers into groups based on their demographic characteristics. Cluster analysis was employed using the variables: age, education and employment status to identify discrete groups of consumers based on their preferences. The results indicate that there are three distinct groups of consumers: young professionals, older-technician and oldest-unemployed. The preference ratings each segment assigns to the three production systems are shown in Table 4. These results show that there were no statistically significant differences among the consumer segments in their preferences for the food production systems. Table 5 indicates priorities each segment assigned to criteria used to assess the food production systems; young professionals accorded a higher priority to community development concerns than the other two groups.

Table 4: Consumer attitudes toward food production systems for each segment

Production Systems	Clusters			Kruskal Wallis Test	
	Young professionals	Older technician	Oldest-unemployed	Chi-Square	Asymp. Sig
Conventional	0.170	0.188	0.211	1.287	0.526
Sustainable	0.274	0.284	0.253	2.264	0.322
Organic	0.556	0.528	0.536	1.242	0.537

Table 5: Consumer attitudes toward the criteria generated by production systems for each segment

Criteria	Clusters			Kruskal Wallis Test	
	Young professionals	Older technician	Oldest-unemployed	Chi-Square	Asymp. Sig
Environmental Concerns	0.142	0.158	0.132	1.783	0.410
Food Safety	0.270	0.281	0.309	4.569	0.102
Food Quality	0.199	0.222	0.208	2.493	0.288
Wellness	0.292	0.261	0.277	1.135	0.567
Community Development Concerns	0.097	0.078	0.073	5.273	0.072

Figure 4 shows the consumers' perceptual map derived from multidimensional scaling. The map illustrates the pattern of proximities for food production systems and the criteria consumers used to assign preference ratings. Kuskal's stress value was used to measure goodness-of-fit. The stress value is a number on a scale from 0 (perfect fit) to 100 (the map captures nothing about the data). In general, researchers are looking for a stress value less than 20 [38]. In the MDS results, Kruskal's stress value is 10 for this two dimensional model and $R^2 = 0.97$. Similar to factor analysis, there is a measure of difficulty in interpreting the conceptual mapping of consumers' perception. To overcome this difficulty, researchers rely on their knowledge of the subject, existing theory and plausible rationale along with the weights associated with the stimulus coordinates to make good sense of the derived stimulus configuration [39]. The results indicate that consumers view organic production systems as quite dissimilar to the other production systems. Additionally, organic production is perceived as being associated with food safety and wellness, but not with environmental and community development benefits. On the other hand, consumers perceive a sustainable system of production to be associated with environmental and community development and food quality. Consumers see conventional as being dissimilar to organic and sustainable production systems and not associated with environment, community development, food quality, food safety and wellness. Consequently, the y axis is labeled as environmental / community development and the x axis as conventional production system. This means that moving from left to right along the x axis the production system becomes more conventional, and moving along the y axis from top to bottom environmental sensitivity of the production system decreases.

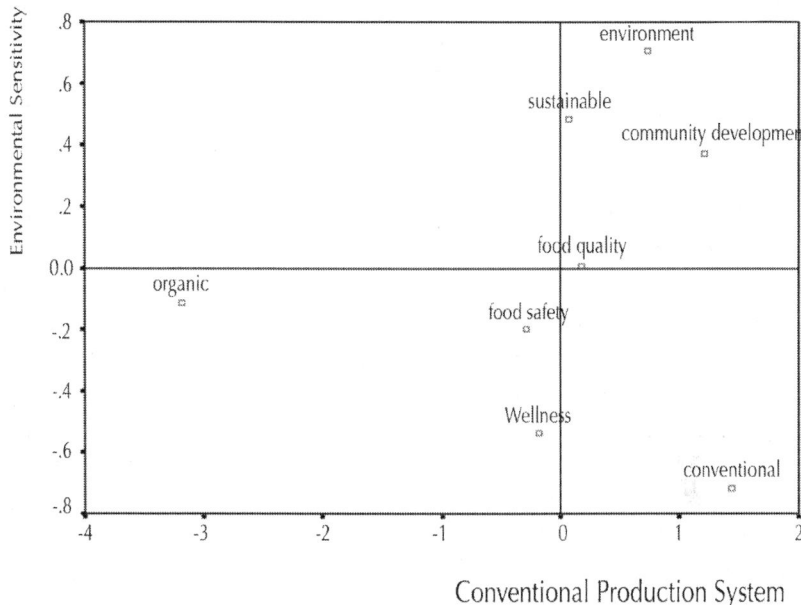

Figure 4: Perceptual mapping of consumers attitudes toward food production systems.

Conclusions

Consumers accord the highest preference score to organic production, followed by sustainable and conventional production systems respectively. Moreover, in according higher priority to food safety and wellness, consumers appear to be more concerned with criteria that are more tangible in terms of the consequence for consumers' personal and immediate well-being. Since our findings indicate that consumers don't associate organic food production with benefits for environmental and community development, there is a need to design education programs that will convince consumers that there are socioeconomic and environmental benefits to be derived from organic production. However, education programs without community institutional support are not likely to succeed. Community members must be engaged as co-creators of initiatives that are intended to change attitudes and create awareness. We recall that proactive and reflective community members live in the same information rich environment as their leaders and those of us considered to be experts. Proactive and reflective citizens tend to assign less significance to leaders and experts, they insist on participating in the decision making process, they want to co-create programs that have implications for their livelihood. As a result, a truly sustainable food system (organic) must become embodied, and an intimate part of the lived experience of people and communities. After all, it is action that creates destiny. So if a sustainable food system is to become a part of our future, it has to become a way of life, and a pattern of living that is acted out as part of the everyday life story of communities [4].

MEASURING PREFERENCES FOR FOOD ATTRIBUTES

Data and Methodology

The sample was designed following the protocol described by [29]. The sample was drawn proportionate to population size by county in Georgia, North Carolina and South Carolina. After specifying the sampling frame parameters, the required sample was purchased from Survey Sampling Inc. Researchers designed and formatted a Fuzzy Pair-wise Comparison (FPC) questionnaire to be compatible with the data collection protocol of Survey Monkey, and trained enumerators to use the questionnaire to collect the data. Enumerators asked consumers to make pair-wise comparisons of five food attributes: nutritional value, hygiene, taste, affordable price and freshness, in order to determine their preference for one attribute over the other. The selected attributes are consistent with the studies which have been done in the U.S. [30]. Data were collected from a random sample of 412 respondents.

In this study, FPC was used to derive a measure of an individual consumer's preferences for fresh fruit and vegetable attributes. The main reasons for using FPC are: 1) The FPC is similar to traditional pair-wise comparisons. Consumers are asked to compare the attributes one pair at a time. However, unlike the traditional pair-wise method, consumers are not forced to make a binary choice between two attributes.

Consumers are permitted to indicate the degree of preference for one attribute over another, and response indicating indifference between attributes is permitted. 2) Unlike the other methods, the scale values are based on the respondent's entire set of paired comparisons. 3) FPC more accurately represents the natural range of response patterns that are possible. The consumer's fuzzy preference matrix R with elements can be constructed as follows [40]:

$$R_{ij} = \begin{cases} 0 & \text{if } i = j \forall i, j = 1,...,n \\ r_{ij} & \text{if } i \neq j \forall i, j = 1,...,n \end{cases} \tag{1}$$

In the FPC method, a measure of preference, μ can be calculated for each attribute by using the consumer's preference matrix R. The intensity of each preference is measured separately using the following equation:

$$\mu_j = 1 - \left(\sum_{i=1}^{n} R^2_{ij} / (n-1) \right)^{1/2} \tag{2}$$

where μj has a range in the closed interval [0,1]. A larger value for μj indicates greater intensity of preference for attribute j. Consequently, fresh fruit and vegetable attributes are ranked from most to least preferable by evaluating the μ values. Then, Friedman and Kendall's W tests were used to evaluate the relative importance of attributes and the extent of agreement among consumers with respect to two or more rankings. In identifying consumer preferences, researchers ranked the importance of the attributes following [37].

Cluster analysis was used to separate consumers into groups using the variables: age, education and employment status. Cluster analysis is a technique used for combining observations or objects (answer, person, opinion, etc.) into groups or clusters. The aim of cluster analysis is to classify observations into relatively homogeneous groups called clusters such that each cluster is as homogeneous as possible with respect to the clustering variables [35-36]. The Kolmogorov-Smirnov normality test was used to check whether the clustering variables showed normal distribution, and then the Kruskal Wallis test was used to compare different groups of clusters.

Multidimensional Scaling (MDS) was used to obtain a perceptual mapping of consumers' preferences for fresh fruit and vegetable attributes. Given a matrix of perceived similarities between attributes of fresh fruit and vegetables, MDS plots the attributes on a map such that those attributes that are perceived to be very similar to each other are placed near each other on the map, and those attributes that are perceived to be very different from each other are placed far away from each other on the map.

Results and Discussion

In this study, consumers were grouped into three clusters. The mean of the variables used in the analysis is presented by clusters in Table 6. There were statistically significant differences among clusters on the variables; age, education and employment of consumers in the sample. The mean age (37.19) is the lowest in Cluster 1 and the highest (77.54) in Cluster 3. Education level is the highest (5.63) in Cluster 1, whereas Cluster 3 has the lowest level (3.87). Employment status changes from employed in Cluster 1 (2.06) to unemployed in Cluster 3 (2.89). Therefore, cluster 1 is labeled "Young professional", while the cluster 2 and cluster 3 are labeled "older-employed" and "oldest-unemployed", respectively.

Table 6: Cluster analysis by age, education and employment

Variables	Clusters			Kruskal Wallis Test	
	1	2	3	Chi-Square	Asymp. Sig
Age	37.19	58.15	77.54	339.960	0.000
Education+	5.63	4.57	3.87	24.101	0.000
Employment++	2.06	2.33	2.89	92.656	0.000

[i] - [+]1: Less than high school, 11: Professional/doctorate degree;+[+]1: Part time, 2: Full time, 3: Unemployed

Table 7 indicates the number of consumers by clusters. The data show that 47.7 percent of consumers are "older-employed worker", whereas 34.8 percent are "young professional", while 17.5 percent represent "oldest-unemployed".

Table 7: Consumer distribution by clusters

Clusters	Frequency	Percent	Cumulative Percent
Young professional	141	34.8	34.8
Older-employed worker	193	47.7	82.5
Oldest-unemployed	71	17.5	100.0
Total	405	100.0	

Descriptive statistics for consumers' pair-wise comparisons of the attributes of fresh fruit and vegetables obtained from the FPC model are presented in Table 8. The fresh fruit and vegetable attributes are ranked from most to least preferable using the reported degree of the consumers' preferences. The results show that the fresh fruit and vegetable attribute most preferred by consumers is freshness with a preference rating of 0.579. Gao, et al. [37] reported a similar pattern of preference in their study on consumer preferences for fresh citrus. Consumers prefer the other food attributes in the following order: taste (0.452), hygiene (0.449), nutritional value (0.428) and affordable price (0.411). In this sample, consumers seem to value freshness, taste and hygiene over price and nutritional value. The Friedman test was used to see if there was a difference in the rankings of the fresh fruit and vegetable attributes.

Table 8: Descriptive statistics of consumer preferences towards fresh fruits and vegetable attributes

Attributes	Mean	Standard deviation	Minimum	Maximum
Nutrition Value	0.428	0.122	0.024	0.929
Hygiene	0.449	0.142	0.049	1.000
Taste	0.452	0.128	0.049	0.868
Affordable Price	0.411	0.154	0.000	0.735
Freshness	0.579	0.159	0.150	1.000

[i] - Significant by Friedman test for $p < 0.01$; Kendall's W=0.11

The Friedman test, which is significant (χ^2=177.71; p<0.01), confirms that some attributes are preferred over the others. Kendall's W test was used to measure the degree of agreement among consumers. The value of Kendall's W is 0.11, which indicates that the level of agreement among consumers in ranking the attributes is very low. A low level of agreement among consumers is an indication of the heterogeneity of consumers' preferences for the attributes of fresh fruits and vegetables.

Since consumers' preferences for the attributes of fruits and vegetables may be influenced by their demographic traits and behaviors [37], demographic traits may be used, where heterogeneity in consumers preferences exists, to segment consumers into groups based on their demographic characteristics. The present study employed cluster analysis using the variables: age, education and employment status to identify discrete groups of consumers based on their preferences. The results indicate that there are three distinct groups of consumers: young professionals, older-employed worker and oldest-unemployed. The results also showed that there was a statistically significant difference among the groups in their preferences for the freshness attribute of fruits and vegetables. Young professionals accorded a higher priority to freshness than the other two groups (Table 9).

Table 9: Consumer preferences for fresh fruits and vegetable attributes by clusters

Variables	Clusters			Kruskal Wallis Test	
	Young professional	Older worker	Oldest-unemployed	Chi-Square	Asymp. Sig
Nutrition Value	0.414	0.434	0.440	2.980	0.225
Hygiene	0.449	0.449	0.452	0.104	0.949
Taste	0.440	0.456	0.473	1.860	0.395
Affordable Price	0.395	0.413	0.436	1.909	0.385
Freshness	0.598	0.579	0.547	6.027	0.049

Figure 5 shows consumers' perceptual map with attribute positioning derived from multidimensional scaling (MDS) analysis of consumers' preferences for the attributes of fresh fruits and vegetables. In the MDS results, Kruskal's STRESS measure is 0.03863. A satisfactory measure should be less than 0.05 for a two dimensional model [39]. $R2=0.99404$ shows that the model's goodness-of-fit is perfect. The analysis indicates that consumers perceive freshness as a distinct food attribute, which is quite separate from taste, hygiene, nutritional value and affordable price. On the other hand, consumers do not seem to perceive hygiene and nutritional value as distinct attributes, that is, consumers tend to accord the same level of priority to hygiene and nutritional value. Similarly, consumers tend to accord the same level of priority to taste and price.

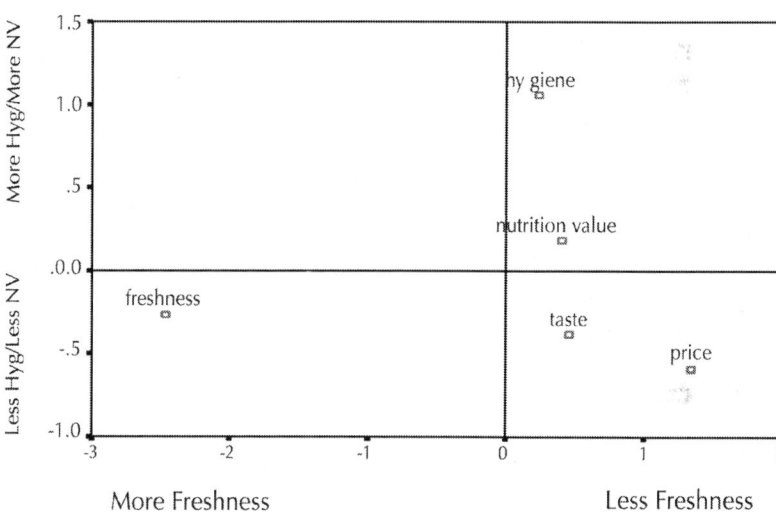

Figure 5: Perceptual mapping of consumers preferences for fresh fruit and vegetable attributes.

CONCLUSIONS

Consumers in making purchasing decisions pay more attention to freshness, taste and hygiene attributes of fresh fruits and vegetables than they do price and nutritional value, when these attributes are considered individually. However, multidimensional scaling shows that consumers tend to associate taste and price when making purchasing decisions, which may explain consumers' love for inexpensive tasty fast food, especially in the case of low income consumers. These results indicate that consumers may not be using all the information available in selecting which food to purchase based on the preference ratings. Therefore, the need exists to educate consumers on the connection among the food attributes and their relevance to healthy eating habits and a healthier lifestyle, particularly the nutritional value attribute. Knowledge about the subgroups of consumers – young professionals, older-employed and oldest-unemployed –provides a basis for farmers, especially farmers supplying urban and suburban farmers' markets, to tailor their products based on the needs of these groups of consumers, a strategy known as market segmentation. For example, the results indicate that the priority or preference of young professionals is for freshness. Extension should use this information to assist farmers to select and display their produce to promote freshness in order to sell more to the higher income young professionals. In summary, these results present extension with an opportunity to (1) assist farmers in marketing their produce in order to meet the needs of specific groups of consumers and (2) in developing a holistic education program, that teaches consumers to use information available on all the attributes: price, taste, hygiene and nutritional value in making purchasing decisions.

In sum, studies have shown that organic farming delivers more environmental benefits, in particular, it delivers more ecosystem services than conventional agriculture [41]. Additionally, contingent on the crop, soil and weather conditions, yield from organic agriculture is equal to that from conventional systems [42]. In the context of a sustainable food production system, organic agriculture goes further in meeting the condition of ecological feasibility (Fig 1), and the evidence seems to indicate that, with further advances in the development of organic technologies, it will become economically feasible. In terms of the third condition to be met-being socially acceptable-in striving for overall sustainability (Fig 1), evidence from our work shows that consumers prefer organic production systems over the alternative systems. Thus far, the future of organic production systems seems promising, but further research is needed to advance the development of organic technologies, disseminate these technologies, increase supply to reduce cost and make organic products affordable to a wider range of consumers, formulate supporting policies, and educate consumers on the value of organic food production systems in contributing to a sustainable food production system.

REFERENCES

1. Wright R.T. Environment Science: Toward a Sustainable Future, 9th Edition. Upper River, NJ: Pearson Prentice Hall; 2005.

2. Wright R.T. Environment Science: Toward a Sustainable Future, 10th Edition. Upper River, NJ: Pearson Prentice Hall; 2008.

3. Kaufmann R. Cleveland C. Environmental Science, 1st Edition. Boston: McGraw Hill; 2008. (Book)

4. Thomas T., Yeboah O., Bukenya J., Gray B., Ofori-Boadu V. Accounting for Socio-Cultural Factors in Designing Sustainable Agricultural Production Systems. Journal of Environmental Monitoring and Restoration 2007; 3(1) 127-139.

5. Thomas T., Yeboah O., Ofori-Boadu V., Fosu E. Assessing Local Community Support for Sustainable Agricultural Production Systems. Journal of Environmental Monitoring and Restoration 2008; 5 191-203.

6. Raven P., Berg L. Environment 5th Edition. NJ: John Wiley & Sons; 2006.

7. Botkin D., Keller E. Environmental Science: Earth as a Living Planet, 5th Edition. NJ: John Wiley & Sons; 2006.

8. Ikerd J. Small Farms: The Foundation for Long-Run Security. Paper Presented at "A Time to ACT: Providing Educators with Resources to Address Small Farm Issues" sponsored by University of Illinois, Agroecology/Sustainable Agriculture Program, Effingham and Peoria, IL; 2002.

9. Carson R. Silent Spring. NY: Fawcett World Library; 1962.

10. Delind L.B. Place, Work, and Civic Agriculture: Fields for Cultivation. Agriculture and Human Values 2002; 19 217-224.

11. Lapping M.B. Big Places, Big Plans. In Furuseth, O. (ed). Perspectives on Rural Policy and Planning. Hampshire: Ashgate Publishing Limited; 2004.

12. Lyson T. A., Guptill A. Commodity Agriculture, Civic Agriculture and the Future of U.S. Farming. Rural Sociology 2004; 69 370-385.

13. Flora C.B. Sustainability of Agriculture and Rural Communities. In Francis, C. A., Flora C.B., King, L.D. (eds). Sustainable Agriculture in Temperate Zones. NY: John Wiley & Sons; 1990.

14. Cone C.A., Myhre A. Community Supported Agriculture: a Sustainable Alternative to Industrial Agriculture? Human Organization 2000; 59(2) 187-196.

15. Lamb G. Community Supported Agriculture: Can it Become the Basis of a New Associative Economy? The Threefold Review 1994; 11 39-44.

16. Rigby D., Caceres D. Organic Farming and the Sustainability of Agricultural Systems. Agricultural Systems 2001; 68 21-40.

17. Horrigan L., Lawrence R.S., Walker P. How Sustainable Agriculture can Address the Environmental and Human Health Harms of Industrial Agriculture. Environmental Health Prospective 2002; 110(5) 445-456.

18. Ikerd J. The Need for a Systems Approach to Sustainable Agriculture. Agriculture, Ecosystems & Environment 1993; 46 147-160.

19. World Commission on Environment and Development. From One Earth to One World: An Overview. Oxford: Oxford University Press; 1987.

20. Snyder, David Pearce. 2007. Five Meta-Trends Changing the World. In Jackson, Robert M. (ed). Annual Editions: Global Issues 06/07. Dubuque, IA: McGraw-Hill; 2007.

21. Borron, S. Building Resilience for an Unpredictable Future: How Organic Agriculture can Help Farmers Adapt to Climate Change. Rome: Food and Agriculture Organization of the United Nations; 2006.

22. Jahn G., Schramm M., Spiller A. The Reliability of Certification: Quality Labels as a Consumer Policy Tool. Journal of Consumer Policy 2005; 28(1) 53–73.

23. Flora C.B., Flora J.L. Entrepreneurial Social Infrastructure: a Necessary Ingredient. The Annuals of the American Academy of Political and Social Science 1993; 529 48-58.

24. Gunden C., Thomas T. Assessing Consumer Attitudes towards Fresh Fruit and Vegetable Attributes. Journal of Food Agriculture & Environment 2012; 10(2) 132-135.

25. Thomas T., Gunden C. Investigating Consumer Attitudes toward Food Produced via Three Production Systems: Conventional, Sustainable and Organic. Journal of Food Agriculture & Environment 2012; 10(2) 132-135.

26. Stewart H., Harris J.M. Obstacles to Overcome in Promoting Dietary Variety: The Case of Vegetables. Review of Agricultural Economics 2004; 27 21-36.

27. U.S. Department of Agriculture. Increasing Fruit and Vegetables Consumption through the USDA Nutrition Assistance Programs. Food and Nutrition Service Progress Report, VA; 2008.

28. Food and Agricultural Organization (FAO) Increasing Fruit and Vegetable Consumption Becomes a Global Priority: 2003.

29. Dillman D., Smyth J., Christian L. Internet, Mail, and Mixed Mode Surveys: The Total Design Method, 3rd Edition. NJ: John Wiley & Sons; 2009.

30. Moser R., Raffaelli R., Thilmany-McFadden D. Consumer Preferences for Fruit and Vegetables with Credence-Based Attributes: A Review. International Food and Agribusiness Management Review 2011; 14 121-142.

31. Saaty T.L. The Analytic Hierarchy Process: Planning, Priority Setting, Resources Allocation. NY: McGraw-Hill; 1980.

32. Saaty T.L. Fundamentals of Decision Making and Priority Theory. Pittsburgh: RWS Publications; 2006.

33. Saaty, T.L. 2008. Decision Making for Leaders. RWS Publications, Pittsburgh.

34. Saaty T.L., Kirti P. Group Decision Making: Drawing Out and Reconciling Differences. Pittsburgh: RWS Publications; 2008.

35. Tabachnick B.G., Fidell L.S. Using Multivariate Statistics, 5th Edition. NY: Pearson; 2007.

36. Hair J.F., Black W.C., Babin B.J., Anderson R.E. Multivariate Data Analysis, 7th Edition. NJ: Prentice Hall; 2010.

37. Goa Z., House L.O., Gmitter F.G., Valim M.F., Plotto A., Baldwin E.A. Consumer Preferences for Fresh Citrus: Impacts of Demographic and Behavioral Characteristics. International Food and Agribusiness Management Review 2011; 14 23-40.

38. Johnson K. Qualitative Methods in Linguistics. Oxford: Blackwell Publishing; 2008.

39. Mazzocchi M. Statistics for Marketing and Consumer Research. London: SAGE Publications; 2008.

40. Van Kooten G.C., Schoney R.A., Hayward K.A. An Alternative Approach to the Evaluation of Goal Hierarchies among Farmers. Western Journal of Agricultural Economics 1986; 11 40-49.

41. Sandhu, H. S., Wratten, S. D., & Cullen R. Organic Agriculture and Ecosystem Services. Environmental Science & Policy 2010; 13 1-7.

42. Pimentel, D., Hepperly, P., Hanson J., Douds, D., & Seidel, R. Environmental, Energetic, and Economic Comparisons of Organic and Conventional Farming Systems. BioScience 2005; 55 (7) 573-582.

Chapter 4

The Suitability of Different Winter and Spring Wheat Varieties for Cultivation in Organic Farming

Beata Feledyn-Szewczyk[1], Jan Kuś[1], Krzysztof Jończyk1, and Jarosław Stalenga[1]

[1]Institute of Soil Science and Plant Cultivation – State Research Institute, Department of Agrosystems and Economics of Crop Production, Puławy, Poland

INTRODUCTION

Nowadays, an interest in organic farming is increasing in many countries [1]. In this specific farming system use of synthetic fertilizers and chemical plant protection measures is forbidden. The limitation of pests, diseases and weeds is provided by agricultural practices which create a beneficial condition of canopy and soil, such as crop rotation, fine and careful tillage, organic fertilization, date and density of sowing as well as mechanical, biological and physical methods of plant protection. An appropriate variety choice is a crucial component of agricultural practices. Selection of cereal varieties suited to organic agriculture requires a different approach to that used in the conventional high input system [2-4].

Organic farming gives a reference to varieties of cereals which are characterized by winter hardiness, high competitive ability against weeds (growth rate, length of stem, tillering rate, surface and angle of leaf attachment), tolerance to fungal diseases, ability to uptake of nutrients as well as resistance to nutrients deficiency stress [5-14]. Moreover the ability of symbiosis with mycorrhizal fungi can be important as it supports uptake of water and mineral nutrients from the soil and improves wholesomeness of plants [4]. Preliminary research indicates organic farming system has low influence on physical and chemical characteristics of grain, with the exception of the total protein content [15].

Some authors suggest that old varieties of cereals were characterized by greater competitiveness against weeds because of longer stems, more prolific tillering and larger leaf area, which increase shading and decrease weed infestation [2, 16, 17]. In breeding process, the stem has become shorter and other features connected with high yields were promoted, according to chemical plant protection [18]. Because of the lower competitiveness of modern varieties against weeds, there can be a problem with selection the varieties to organic and other less intensive crop production systems.

Christensen's study [7] showed no significant correlation between yield and competitive ability of cereal varieties against weeds, which means that both of these features should be considered by breeders in selecting varieties for organic farming system. Promising might be the selection for genotypes with high early nitrogen uptake efficiency amongst those already recognised as having good coverage and shading ability [4].

The aim of the research was to compare the yields of several winter and spring wheat varieties cultivated in organic and conventional crop production systems to identify the causes of yield differentiation in these systems.

MATERIAL AND METHODS

The study was conducted in 2008-2010 at the Experimental Station of Institute of Soil Science and Plant Cultivation – State Research Institute in Puławy, Poland (N:51°28′, E:22°04) in an experiment, in which different crop production systems have been compared since 1994. The experiment was located on a Luvisol, a loamy sand and on a sandy loam. The characteristics of agricultural practices used in each system were presented in Table 1.

Table 1: Major elements of the agricultural practices of winter wheat and spring wheat cultivated in organic and conventional farming systems

	Farming system	
	Organic	Conventional
Crop rotation	Potato spring wheat ± undersown crop clover and grasses (1 year) clover and grasses (2 year) winter wheat ± catch crop	Winter rape winter wheat spring wheat
Seed dressing	-	+
Organic fertilization	compost (30 t·ha-1) under potatoe + catch crop	winter rape straw, winter wheat straw
Mineral fertilization (kg·ha-1)	based on soil testing, allowed P and K fertilizers in form of natural rock were used: 150 kg of potassium sulphate (75 kg K2O), phosphate rock powder 150 kg (42 kg P2O5)	winter wheat: NPK (140+60+80) spring wheat: NPK (70+60+45)
Fungicide	-	winter wheat: 2 – 3 x spring wheat: 1 x

Growth regulators	-	winter wheat: 2 x spring wheat: 1 x
Weed control	winter wheat: harrowing 2-3 x spring wheat: none (undersown clovers + grasess)	winter wheat: herbicide 2-3 x spring wheat: harrowing 1x + herbicide 1x

The area of each crop rotation field was 1 ha. Within the field of winter and spring wheat, the experiment with different varieties was established in completely randomized blocks. Five modern varieties of common wheat (*Triticum aestivum ssp. vulgare*): Kobra Plus, Bogatka, Smuga, Tonacja, Ostka Strzelecka and old variety of spelt wheat (*Triticum aestivum ssp. spelta*)-Schwabenkorn were cultivated. Nine spring wheat varieties: Bombona, Vinjett, Parabola, Tybalt, Nawra, Raweta, Bryza, Żura, Zadra were tested in organic system. In conventional system 4 varieties of winter wheat: Kobra Plus, Bogatka, Rywalka, Legenda and 4 varieties of spring wheat: Bombona, Parabola, Tybalt and Vinjett were cultivated. The varieties of common wheat were sown at a rate of 220 kg grains · ha^{-1}, spelt wheat - 200 kg of spikelets· ha^{-1}, spring wheat – 190 kg grains · ha^{-1}. In conventional system spring wheat was cultivated in pure stand whereas in organic system it was sown with undersown crop (common clover – 10 kg ha^{-1}+Dutch clover 3 kg ha^{-1}+meadow fescue 10 kg ha^{-1}+perennial ryegrass 10 kg ha^{-1}). According to organic agriculture rules, mineral nitrogen fertilizers and chemicals were not used. In winter wheat canopy weeds were controlled in mechanical way, using a weeder, two or three times in spring, at tillering stage, whereas in spring wheat mechanical weed control was not applied due to undersown plants. In conventional farming system chemical plant protection was applied (Table 1).

High differences in weather conditions were observed during the research period (Table 2). The growing season of 2007/2008 was characterized by favourable weather conditions for growth and development of winter wheat and spring wheat. Precipitation slightly higher than average was well distributed. Suitable thermic and moisture conditions reflected in the development of dense wheat canopies of large ability to compete with weeds. The spring of 2009 was delayed, which limited the effectiveness of harrowing. Threefold harrowing of winter wheat caused damage to plants. Late spring, lack of precipitation and night frosts until mid-May disturbed the phases of tillering, stem elongation and earing. As a result, the density of wheat canopies was low, which created favorable conditions for weeds. The weather conditions in the autumn of 2009 were suitable for growth and development of winter wheat. Snow mould and frost till the end of April of 2010 had adverse effects on development of winter wheat. Sparse wheat canopies competed worse with weeds. Growing season was characterized by an unfavorable distribution of temperature and precipitation: dry April and heavy rains in May (110 mm). High temperature and drought in June and July negatively affected yields of spring wheat (Table 2).

Table 2: Mean daily temperature of air (°C) and monthly sum of precipitation (mm) in Osiny in 2007-2010 compared to the longtime average (1951-2006)

Months	Temperature (°C)				Precipitation (mm)			
	2007/2008	2008/2009	2009/2010	1951-2006	2007/2008	2008/2009	2009/2010	1951-2006
IX	13.1	12.5	14.8	13.2	86	61	22	49
X	7.8	9.9	6.8	8.2	7	42	77	44
XI	1.4	5.1	5.3	2.7	36	20	42	39
XII	-0.7	1.3	-1.7	-1.4	6	35	42	37
I	1.0	-2.6	-8.3	-3.4	46	23	22	31
II	2.9	-0.7	-1.7	-2.4	11	29	29	29
III	3.9	2.2	3.0	1.7	39	61	13	30
IV	9.5	11.0	9.3	7.9	43	2	17	40
V	13.5	13.7	14.3	13.5	83	63	110	57
VI	18.2	16.6	18.3	16.8	42	96	48	70
VII	18.8	20.1	22.1	18.5	94	69	43	85
VIII	18.6	18.4	20.3	17.3	72	98	119	75

Grain yield and components of its structure were estimated in 25m² plots in 8 replications. Stem and leaves were scored for infestation rate with fungal pathogens at mik-dough stage (BBCH 77-83). Forty individuals in 4 replications for each variety were taken for plant pathology analysis purposes. Percentage of the disease-damaged leaf blade surface was determined in accordance with the recommendations of the EPPO [19]. A 4-step infestation scale was used to calculate the stem base infection index.

The number of weeds and their dry matter were assessed at dough stage of wheat (BBCH 85-87), on an area of 0.5 m², in four replications for each variety. Moreover, the biometric analysis of wheat plants and plant canopy, such as plant height, overall tillering, number of plants and dry matter of wheat per unit area (1 m) were done.

Total protein of grain was determined based on the content of nitrogen measured by Kjeldahl Nitrogen Determination Method x 5.83.

The results were analysed statistically. Analysis of variation for the completely randomized model was applied and the significance of differences between means was verified by Tukey test at α=0.05. Pearsons' correlations between grain yield of winter and spring wheat varieties in organic system and yield limiting factors as well as morphological features and canopy parameters were performed. Cluster analysis using Furthest Neighbour method was done in order to divide varieties into groups with similar characteristics. Calculations were performed using Statgraphic Plus version 2.1.

RESULTS AND DISCUSSION

Estimation of Winter Wheat Varieties Suitability for Cultivation in Organic Farming

Yields of winter wheat cultivated in organic farming system were strongly differentiated in years. An average for all varieties reached from 2.1 t ha⁻¹ in 2010 (unfavorable weather conditions) to 5.4 t ha⁻¹ in 2008 (good weather conditions) (Table 3, 4).

No clear information about yields of winter wheat varieties cultivated in organic farming was achieved. Among the varieties that are in the

Registry, Smuga gave the highest yields. That variety yielded so high in the growing season of 2009/2010 thanks to its good winter survival and high plant density (table 4). Varieties that yielded the lowest were: Ostka Strzelecka and spelt Schwabenkorn – 3.2 t ha^{-1} (Table 3).

Grain yield in conventional farming system was, on average of 4 varieties, 45% higher than in organic farming system (Table 3, 5). The difference between those two farming systems was: 30% in 2008, 49% in 2009 and 61% in disaster-year 2010. Yield differences between organic and conventional farming systems were 37% for a cv. Kobra and 45% for a cv. Bogatka. Research conducted in 2005-2007 on varieties: Roma, Kobra, Zyta, Sukces revealed that average yield differences between organic and conventional farming system were 19%. The old varieties of common wheat: Ostka Kazimierska, Kujawianka Więcławicka, Wysokolitewka Sztywnosłoma were not very useful for cultivation in organic system because of low grain yield [20]. Number of ears per 1 m^2 in organic system was lower than in conventional one, despite the same sowing rate, by 14% in 2008 to 47% in 2010, due to lower density of wheat in organic farming system after the severe winter. Also, the differences between grain weight between two farming system were observed (15% for Kobra and Bogatka cultivars on average) (table 3, 5).

Table 3: The grain yield of winter wheat and the elements of yield in organic system

Cultivar	The yield of grain (t·ha-1)				Number of ears (pcs·m-2)				1000 grains weight (g)			
	2008	2009	2010	mean	2008	2009	2010	mean	2008	2009	2010	mean
Kobra Plus	6.18	2.92	2.39	3.83	569	323	268	387	46.8	32.0	38.0	38.9
Bogatka	6.21	2.33	2.36	3.63	479	278	242	333	50.9	38.9	44.3	44.7
Smuga	5.82	3.91	3.04	4.26	576	392	320	429	47.2	36.4	38.6	40.7
Tonacja	5.56	3.53	1.80	3.63	488	444	151	361	40.1	37.2	39.6	38.9
Ostka Strzelecka	4.94	3.48	1.09	3.17	457	423	165	348	38.0	33.1	34.0	35.0
Spelt*	3.80	3.96	1.94	3.23	562	444	251	419	76.1	76.6	76.0	76.2
Mean	5.42	3.36	2.10	3.63	522	384	233	380	without spelt 44.6	35.5	38.9	39.6
HSD (=0.05)	0.33	0.28	0.31		51	51	66		2.1	2.9	2.0	

[i] - * glume grains and the 1000 spikelets weights

Table 4: Survival of winter wheat varieties during winter 2009/2010 in organic system

cultivar	% loss of plants during the winter
Kobra Plus	33
Bogatka	41
Smuga	17
Tonacja	45
Ostka Strzelecka	30
Spelt	35

Table 5: The grain yield of winter wheat grain and the elements of yield in conventional system

Cultivar	The yield of grain (t·ha-1)				Number of ears (pcs·m-2)				1000 grains weight (g)			
	2008	2009	2010	mean	2008	2009	2010	mean	2008	2009	2010	mean
Kobra Plus	7.41	6.47	4.45	6.11	663	483	501	549	49.6	44.6	43.5	45.9
Bogatka	7.82	6.73	5.36	6.63	539	447	419	468	56.6	50.1	51.5	52.7
Rywalka	8.18	6.41	6.11	6.90	614	458	409	494	54.1	44.7	48.7	49.1
Legenda	7.73	6.81	5.65	6.73	607	457	432	499	50.7	42.2	49.3	47.4
Mean	7.78	6.60	5.39	6.59	606	461	440	502	52.9	45.4	48.3	48.8
HSD (α=0.05)	0.51	ns*	0.31		63	ns	36		1.2	0.9	1.3	

[i] - * ns – non significant differences

According to Seufert et al. [21], crop yields in organic farming are from 5 to 34% lower than those in conventional farming. Those differences depend on plant species, soil type, fertilization, agriculture level and economic development of the country. Tyburski and Rychcik's [22] study showed that winter wheat yield can reach 4.27 t ha^{-1} in an organic farm and 5.63 t ha^{-1} in a conventional one. According to the authors correct crop management in some of organic farms can result in crop yields to be almost as high as in conventional farms.

Weed infestation is one of the most strongly limiting factors to cereals yields in organic farming system. Weed communities differed in years and between tested varieties. The lowest level of weed infestation was observed in 2008 (19-46 g m^{-2}), whereas the biggest dry matter of weeds was noted in 2010 (170-345 g m^{-2}) (Table 6), the latter being connected with sparse canopies after winter (Table 4). Percentage loss of plants during the winter was the highest in Bogatka and Tonacja and the lowest in Smuga (Table 4). The comparison of common wheat and spelt wheat varieties showed the highest abundance of weeds in Tonacja, Kobra Plus and Bogatka (Table 6). Dry matter of weeds was the largest in Bogatka and Kobra Plus canopies. The lowest number of weeds at dough stage was recorded in Ostka Strzelecka and spelt Schwabenkon and the lowest dry matter of weeds in Smuga. In canopies of all varieties a few dicotyledonous species dominated: *Viola arvensis, Stellaria media, Papaver rhoeas, Polygonum convolvulus, Chenopodium album*. Monocotyledonous species were represented mainly by *Apera spica-venti*.

Weed infestation of winter wheat cultivated in conventional system was low (3.0 plants m^{-2}, 2.1 g m^{-2}on average) and did not differ significantly between varieties and years due to the intensive chemical weed control using herbicides (Table 1, 7). That level of weed abundance did not affect yields of winter wheat.

Varieties of common wheat and spelt wheat differed for some morphological features and for canopy structure. Among tested cultivars spelt Schwabenkorn was significantly the tallest (122 cm on average at dough stage) (Figure 1). Differences in height between spelt and common wheat varieties increased with advancing plant age, which influenced the competitive ability against weeds and was reflected in dry matter of weeds (Table 6). The lowest variety in the first stages of growth was Tonacja, while at dough stage Kobra Plus characterized with significantly the smallest stem length (85 cm on average) (Figure 1). In both of these varieties high level of infestation was indicated (Table 6). Among the varieties of common wheat Smuga was the highest in all phases of development and it was the most competitive against weeds.

Table 6: Number and dry mater of weeds in winter wheat cultivated in organic system as determined at dough stage

Cultivar (A)	Number of weeds (plants·m-2)				Dry matter of weeds (g·m-2)			
	Years (B)				Years (B)			
	2008	2009	2010	mean	2008	2009	2010	mean
Kobra Plus	45.0	147.5	117.5	103.3	42.8	188.3	225.7	152.3
Bogatka	57.0	87.5	114.0	86.2	45.6	184.2	344.5	191.4
Smuga	61.0	79.5	101.0	80.5	29.7	104.8	169.3	101.3
Tonacja	43.0	156.5	132.5	110.7	18.3	84.2	293.5	132.0
Ostka Strzelecka	64.5	73.0	81.0	72.8	42.3	91.0	291.1	141.5
Spelt	67.0	91.0	68.5	75.5	30.0	114.9	224.0	123.0
Mean	56.3	105.8	102.4	88.2	34.8	127.9	258.0	140.2
HSD (α=0.05) for:	A - 34.9, B - 20.1, interaction AB - 57.8				A - 63.9, B - 36.9, interaction AB - 106.1			

Table 7: Number and dry mater of weeds in winter wheat cultivated in conventional system as determined at dough stage

Cultivar (A)	Number of weeds (plants·m-2)				Weeds dry matter (g·m-2)			
	Years (B)				Years (B)			
	2008	2009	2010	mean	2008	2009	2010	mean
Kobra Plus	3.0	5.0	1.0	3.0	1.2	3.8	0.8	1.9
Bogatka	2.5	4.5	0	2.3	3.9	1.2	0	1.7
Rywalka	4.5	4.0	6.0	4.8	10.0	2.1	0.6	4.2
Legenda	1.0	3.5	1.0	1.8	0.4	0.6	0.1	0.4
Mean	2.8	4.3	2.0	3.0	3.9	1.9	0.4	2.1
HSD (α=0.05) for:	A - ns*, B - ns, interaction AB - ns				A – ns, B - ns, interaction AB - ns			

[i] - * ns – non significant differences

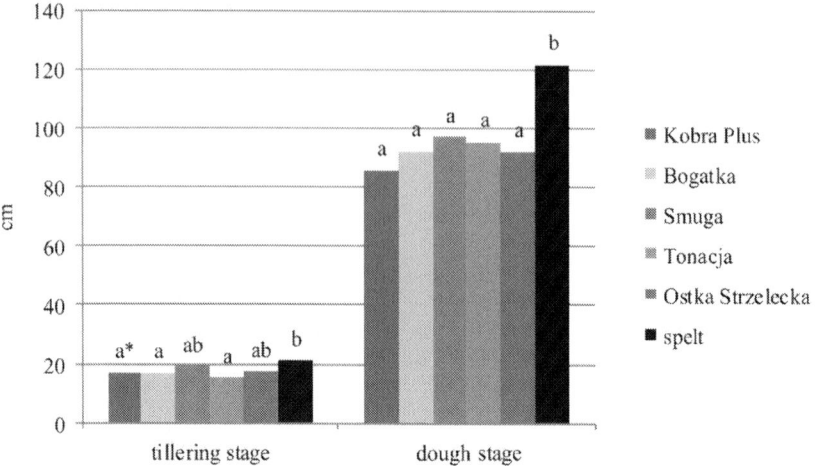

Figure 1: Height of winter wheat varieties cultivated in organic system (mean from 2008-2010).

* data marked with the same letters do not differ significantly at α=0.05

Spelt Schwabenkorn was characterized by significantly the most profuse overall tillering (Figure 2). There were no differences in tillering of modern varieties of winter wheat.

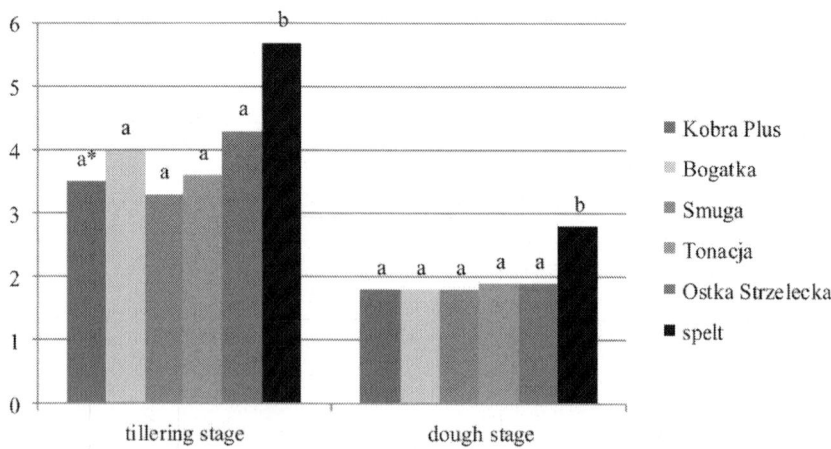

Figure 2: Tillering of winter wheat varieties cultivated in organic system (mean from 2008-2010).

* data marked with the same letters do not differ significantly at α=0.05

The highest density of plants was noted in a Smuga throughout the growing season and the lowest in spelt Schwabenkorn (Table 8). Plant density was the highest in the best weather conditions of 2008 and lowest in 2010, which was caused by poor winter survival of wheat (table 1, 4, 8). In 2010 weed infestation was determined more by plant density than by morphological features of varieties. The largest biomass was produced by Smuga (Table 8) which could be the reason for its big competitiveness against weeds (Table 6). Bogatka and Kobra Plus canopies were characterized by the lowest biomass.

Table 8: Selected features of canopy of winter wheat varieties cultivated in organic system

Cultivar (A)	Density of wheat plants (plants·m-2)				Dry matter of wheat (g·m-2)			
	Years (B)				Years (B)			
	2008	2009	2010	mean	2008	2009	2010	mean
Kobra Plus	271	214	116	200	684	713	657	685
Bogatka	238	223	102	188	738	610	651	666
Smuga	291	242	156	230	817	860	848	842
Tonacja	246	219	115	193	715	809	590	705
Ostka Strzelecka	253	239	122	205	707	893	536	712
Spelt	234	192	123	183	746	976	680	801
Mean	256	222	122	200	735	810	660	735
HSD (=0.05) for:	A – 42.1, B – 24.2, interaction AB – ns*				A – 137.6, B – 79.2, interaction AB – 228.0			

[i] - * ns – non significant differences

Competitiveness against weeds is one of the criteria of variety selection for the organic farming system [7, 13]. The varieties with a rapid growth rate at the initial growth stages, the biggest height, tillering, the most horizontal (planophile) set of leaves to the soil surface, combined with a low susceptibility to the disease, which prolongs the duration of the foliage, have the highest competitive abilities [2, 8]. These features affect the ability of shading the soil surface, and thus photosynthetically active radiation penetrates into the canopy, which directly influences the growth of weeds. Some authors suggest that the height of the plants is the main reason for the differences in competitiveness against weeds [17, 23, 24], but others believe that this factor has a marginal significance [25, 26]. Recent research on wheat and barley varieties showed that the differences between plant density influence the competitive ability more than do plant height and light penetration in the canopy [27]. Furthermore, some authors suggest that differences between varieties arise from their different allelopathic properties [10, 16]. Moreover, weed infestation suppressing ability of varieties vary because of their combined allelopathic and competitive abilities [28]. Those authors indicates that both the allelopathic potential and the competitiveness connected with morphological characteristics of variety are quantitative factors of complex inheritance which are also dependent on environmental conditions.

The relationship between weed infestation and grain yield of winter wheat varieties was analyzed (Figure 3-5). In 2008 the dry matter of weeds was low, less than 50 g m^{-2} and it probably had no significant effect on grain yield of wheat cultivars. Due to favorable weather conditions of that year grain yields were high, from 3.8 t ha^{-1} of hulled grain of spelt to 6.2 t ha^{-1} in Bogatka and Kobra Plus. At low weed pressure, Kobra Plus and Bogatka gave the highest yield, but when the weed infestation was high, 180 g m^{-2} and more, as in 2009 and 2010, the grain yield was low (Figure 4-5). At the weed infestation of 100-110 g m^{-2} spelt Schwabenkorn and Smuga gave high yields, about 4 t ha^{-1} (Figure 4). In 2010, due to poor winter survival of varieties and unsuitable weather conditions, the yields of all varieties were low (Figure 5). Analysis showed that with the increase in dry weight of weeds grain yields decreased. The highest grain yields and the lowest infestation was found in Smuga, whereas Ostka Strzelecka and Tonacja were the lowest yielding variety with the highest weed weight (280-290 g m^{-2}) (Figure 5).

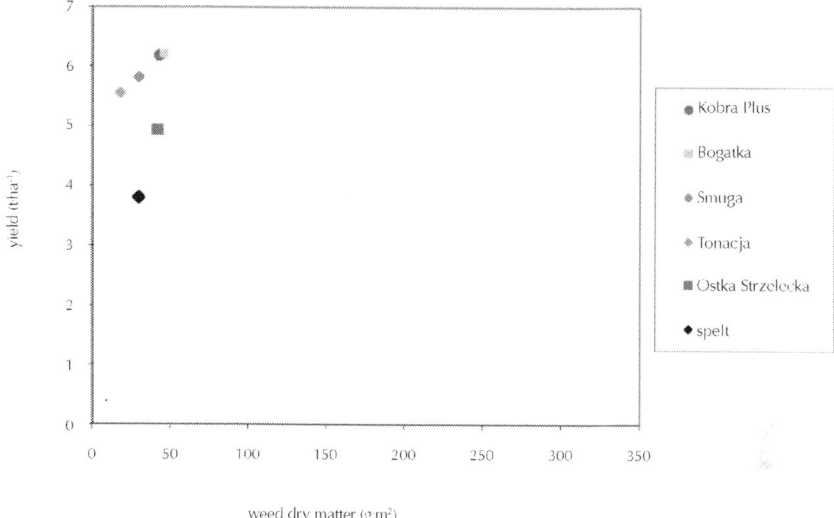

Figure 3: The dependence between weed dry matter and grain yield of winter wheat varieties cultivated in organic system at dough stage in 2008.

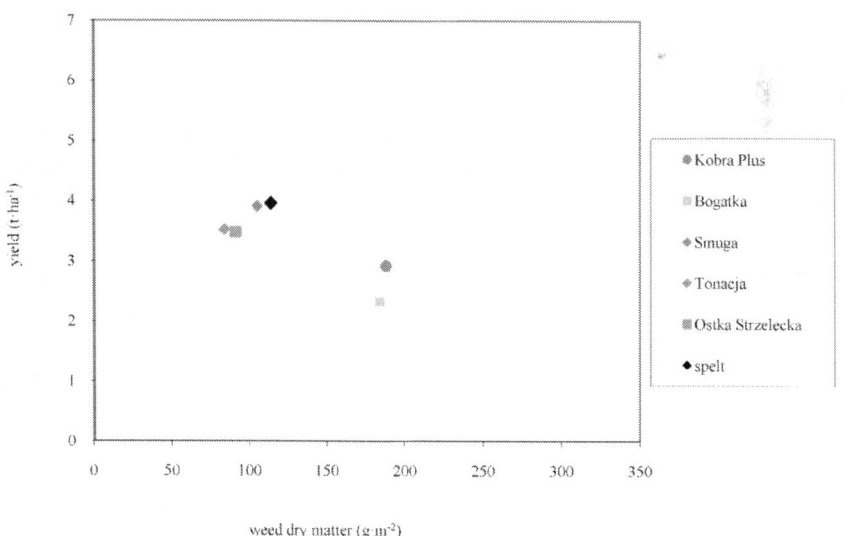

Figure 4: The dependence between weed dry matter and grain yield of winter wheat varieties cultivated in organic system in dough stage in 2009.

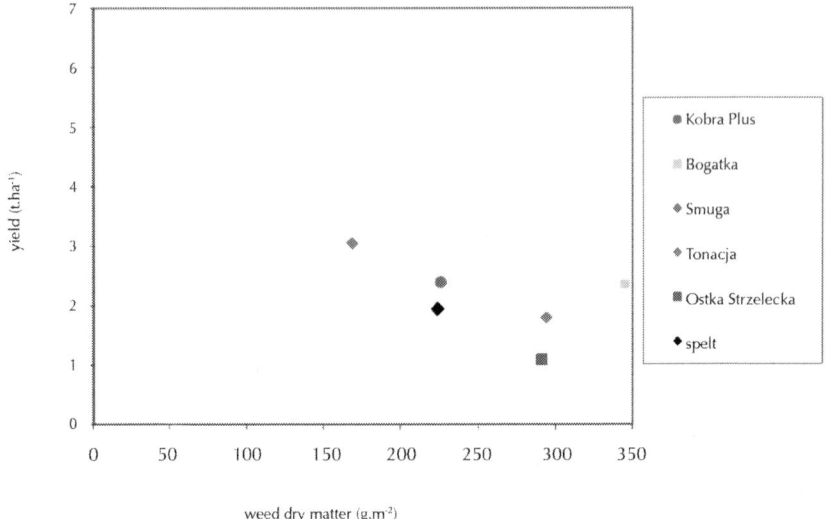

Figure 5: The dependence between weed dry matter and grain yield of winter wheat varieties cultivated in organic system at dough stage in 2010.

The infection by pathogens can be as well as weed infestation the yield limiting factors in organic farming [6]. The similar effect can be produced by restricted nitrogen supply of plants [12]. The results showed that stem base disease incidence was relatively low in winter wheat varieties (approximately 10%) (Table 9). Comparison of sanitary conditions of winter wheat growth under conventional vs. organic system indicate that crop rotation applied in organic farming system effectively limited stem base disease development. Incidence of stem base disease observed at milk-dough stage of winter wheat was twice lower in organic than in conventional farming system, although in conventional system with simplified crop rotation fungicides were used. On the other hand, incidence of leaf diseases was approximately four times higher in organic farming system than in conventional one (Table 9).

The rate of infection by fungal stem base and leaf diseases differed from variety to variety (Table 9). Cvs. Tonacja and Ostka Strzelecka had the lowest rate of stem base disease infection. However those varieties (as well as Smuga) had the highest rate of infection by leaf diseases. A pathogen species infecting the common wheat was also different in years. *Puccinia recodita* had the highest rate of infection in 2008,

Septoria spp. in 2009 and in year 2010 both those pathogens infected common wheat. Cvs. Kobra Plus and Bogatka showed the lowest rate of the leaf disease infestation in organic farming system through the years. However, they were highly infected by stem base diseases (Table 9). Those varieties also revealed a high rate of fungal disease infestation in conventional farming system.

Table 9: The infestation of stem base and leaves of winter wheat by pathogens in milk-dough stage in organic and conventional farming system

Cultivar	Stem base infection index (%)				Leaves infection index (%)*			
	2008	2009	2010	mean	2008	2009	2010	mean
organic system								
Kobra Plus	15.1	12.0	10.4	12.5	4.6	15.0	31.2	16.9
Bogatka	25.1	5.1	10.0	13.4	10.8	11.6	29.7	17.4
Smuga	18.7	10.0	5.6	11.4	21.2	39.7	21.4	27.4
Tonacja	15.8	1.1	5.3	7.4	21.9	22.3	31.1	25.1
Ostka Strzelecka	14.2	1.8	6.3	7.4	20.2	30.3	28.8	26.4
spelt	11.1	8.2	8.3	9.2	18.9	27.3	22.7	23.0
mean	16.7	6.4	7.7	10.2	16.3	24.4	27.5	22.7
conventional system								
Kobra Plus	42.8	2.6	23.7	23.0	1.6	10.9	18.9	10.4
Bogatka	45.2	3.4	32.2	26.9	0.8	11.1	5.2	5.7
Rywalka	34.4	0.5	18.6	17.8	0.4	5.0	0.5	2.0
Legenda	16.4	3.0	20.2	13.2	0.3	10.1	0.5	3.6
mean	34.7	2.4	23.7	20.3	0.8	9.3	6.3	5.4

[i] - *The area (%) of flag and underflag leaf infested by pathogens

Higher leaf infestation by fungal pathogens in organic farming system, as well as nitrogen deficiency during shaping of grains, negatively influenced nitrogen availability and decreased total protein content of the grain (Table 10). Varieties listed in the regional allocation of crops contained an average of 12.6% protein, although protein content in highly fertilized conventional farming was on average 2.1% higher. Differences in the total protein content between varieties were insignificant, both in organic and conventional farming system. Spelt Schwabenkorn was the only exception as it had a clearly higher content of proteins (compared to modern varieties) (Table 10).

Table 10: The content of total protein of the grain (%) of selected winter wheat cultivars in organic and conventional farming system

Cultivar	Years of research and farming systems		
	2008	2009	mean
organic system			
Kobra	11.2	13.6	12.4
Bogatka	11.4	13.6	12.5
Smuga	11.4	14.1	12.8
Tonacja	11.7	13.3	12.5
Ostka Strzelecka	11.7	13.9	12.4
Spelt	16.8	16.3	16.6
mean (without spelt)	11.5	13.7	12.6
conventional system			
Kobra	13.4	15.7	14.6
Bogatka	12.9	15.9	14.4
Rywalka	13.6	16.2	14.9
Legenda	14.0	15.8	14.9
mean	13.5	15.9	14.7

Analysis of correlation was done to assess the relationships between grain yield of winter wheat in organic system and factors limiting the yield as well as morphological features and canopy parameters. This analysis showed strong, negative correlation between grain yield of winter wheat and weed dry matter at dough stage (r=-0.792) (Table 11). The grain yield was also significantly influenced by number of weeds and leaf infestation by fungal pathogens. There was no correlation between stem base diseases and grain yield of winter wheat in organic system. The tillering and height of common wheat did not have a significant impact on yields, but there was a strong correlation between grain yield and plant density. Weed infestation was significantly influenced by density of wheat, dry matter of wheat and its height. There was no significant correlation between tillering and parameters of weed infestation (Table 11).

Table 11: Correlation coefficients (r) between grain yield and factors limited the yield, some morphological features and canopy parameters for all tested varieties cultivated in organic system at dough stage (N=72)

Parameters	Grain yield	Stem base diseases	Leaves diseases	Number of weeds	Dry matter of weeds	Tillering	Height	Density of wheat	Dry matter of wheat
Grain yield	x	0.069	-0.477*	-0.463*	-0.792*	-0.205	0.124	0.716*	0.297*
Stem base diseases		x	-0.434	-0.435*	-0.412*	-0.048	0.272*	0.358*	-0.030
Leaves diseases			x	0.223	0.369*	0.195	-0.116	-0.368*	0.129
Number of weeds				x	0.494 *	-0.099	-0.407*	-0.357*	-0.123
Dry matter of weeds					x	0.103	-0.242*	-0.701*	-0.435 *
Tillering						x	0.597*	-0.366*	0.167
Height							x	0.030	0.188
Density of wheat								x	0.359*
Dry matter of wheat									x

[i] - * significant at α=0.05

The cluster analysis based on a grain yield, weed and fungal infestation, some morphological features and canopy structure at dough stage allowed to divide varieties into 3 groups with similar characteristics (Table 12).

Table 12: The results of cluster analysis based on grain yield, factors limiting the yield, some morphological features and canopy parameters for tested varieties grown in organic system

Cluster	Grain yield (t·ha-1)	Leaves infestation index (%)	Number of weeds (pcs·m-2)	Dry matter of weeds (g·m-2)	Density of wheat stand (pcs·m-2)	Dry matter of wheat (g·m-2)	Varieties
1	3.69	19.8	100.1	158.5	191.6	683.2	Kobra Plus, Bogatka, Tonacja
2	3.20	24.7	74.2	132.2	193.8	756.5	Spelt Schwabenkorn, Ostka Strzelecka
3	4.26	27.4	80.5	101.3	229.7	841.3	Smuga

The first cluster was characterized by the highest level of weed infestation (number and dry matter) and the smallest density of wheat plants and dry matter of wheat. This cluster was represented by 3 varieties: Kobra Plus, Bogatka, Tonacja with the lowest competitive ability against weeds, average level of grain yield and low leaf infestation by fungal pathogens (Table 12). The second cluster grouped 2 varieties: spelt Schwabenkorn and Ostka Strzelecka, with the lowest grain yield, small wheat density, medium level of infestation by weeds and leaf diseases. Third cluster was represented only by one variety – Smuga, which was characterized by the highest yields and the highest stand density and biomass. These parameters of Smuga's canopy influence its competitiveness against weeds and is reflected in the lowest weed infestation. If not for its highest leaves infestation index, Smuga would have been recommended variety for organic system (Table 12).

The most required features of cereal varieties cultivated in organic system are: winter hardiness, high competitive ability against weeds, tolerance of fungal diseases and ability to take up and effectively use fertilizers from soil [5-14]. Smuga, due to its winter hardiness, competitiveness ability against weeds and high yields seems to be suitable for organic farming system. A little less suitable (considering the same features) are spelt Schwabenkorn and Ostka Strzelecka. Kobra Plus, Bogatka and Tonacja were the highest yielders in years with good weather conditions, with optimal plant density and poorer weed pressure. The grain yield of these varieties was low when there was a high level of weed infestation (180 g m^{-2} and more), as in 2009 and 2010. Kobra Plus and Bogatka was also the less leaf fungal-infested varieties in organic system in the research period 2008-2010.

Summary

- In organic farming system, among the tested varieties, Smuga gave the highest yields (4.26 t ha^{-1}), while Ostka Strzelecka was the lowest yielder (3.17 t ha^{-1}). Yields of spelt Schwabenkorn were approximately 13% lower than those of modern varieties (glume grain) (3, 20 t ha^{-1}). Old varieties of common wheat: Ostka Kazimierska, Kujawianka Więcławicka, Wysokolitewka Sztywnosłoma were not very useful for cultivation in organic system because of low grain yield and high infestation of leaves.

- Yields of winter wheat cultivated in conventional farming system were on average 45% higher than those produced in organic system. Research conducted in 2005-2007 on varieties: Roma, Kobra, Zyta, Sukces revealed that average yield differences between organic and conventional farming system were 19%. Lower yields in organic system, despite the same sowing rate, were caused by lower number of spikes per m² (25% on average) and slightly lower grains weight (15%).

- Weed communities differed in years and between tested varieties. The lowest level of weed infestation was observed in 2008 (19-46 g m⁻²), whereas the biggest dry matter of weeds was noted in 2010 (170-345 g m⁻²), which was related to sparse canopies after winter. The comparison of common wheat and spelt wheat varieties showed the highest level of weed infestation in Kobra Plus, Bogatka and Tonacja canopies. The lowest number of weeds at dough stage was shown by Smuga, Ostka Strzelecka and spelt Schwabenkon. Weed infestation was significantly influenced by wheat plant density, dry matter and plant height.

- Stem base diseases has a lower importance in organic farming system than have leaf diseases. Kobra Plus and Bogatka showed the highest resistance to leaf fungal diseases.

- Higher infestation rates by leaf diseases and lower nitrogen supply to the crop under the organic system caused a decrease of total grain protein content compared to that under conventional system. Spelt Schwabenkorn produced grains with the highest protein content.

Estimation of Spring Wheat Varieties Suitability for Cultivation in Organic Farming

In organic farms spring wheat is much more popular than winter wheat. The reason for that is that weed infestation in spring wheat is easier to control than in winter wheat, also fungal diseases in spring wheat are less common. Moreover, it is sown after late-harvested proceeding crops (vegetables, potatoes, fodder crops and intercrops). Spring wheat is also a valuable protective crop for undersown papilionaceous plants and grasses [29].

In organic system the yield of spring wheat, mean for 9 tested varieties, ranged from 4.7 t ha^{-1} in 2008, when the weather was favourable for plant growth, to 2.8 and 3.1 t ha^{-1} in other years (Table 13). The higher yield of spring wheat in 2008 than in the other two years of the study could be caused by higher rainfall in autumn and winter. The sum of rainfall for 6 months of 2007/2008 (X 2007 – III 2008) reached 146 mm, whereas in the subsequent years it reached approximately 220 mm (Table 2). This could cause leaking of mineral nitrogen into the deeper layers of soil resulting in an impaired nitrogen supply to plants. The effects of deficiency became especially manifest in the springtime period of rainfall shortage, when intensity of biological processes slowed down. Such weather conditions prevailed in 2009 and 2010.

Cvs. Nawra and Bryza had the lowest yields while Tybalt, Żura, Zadra and Vinjett were the highest yielders (Table 13). Among those varieties Żura showed the most stable yields (3.31 – 4.86 t ha^{-1}), whereas Tybalt's yields were the most differentiated (6.16 – 2.43 t ha^{-1}). An extremely high plant density as well as grains weight was the main reason of high wheat yields in 2008. Likewise the weed infestation in this year was very low (Table 15).

During the 3-year period of research, yields of spring wheat cultivated in organic farming system were on average lower by 1.86 t ha^{-1} (34%) than those in conventional farming system (Table 13-14). The difference in yields from organic vs conventional system across the years ranged from 30% in 2008 (high yields) to 36% in 2009 (very low yields). Yield comparison between organic and conventional system showed that Parabola's yield was the most influenced by the farming system (38%) while the yields of the other varieties differed from 32 to 33%.

A lower yields in the organic farming system was caused by lower ear density and inferior grain weight (Table 13-14). Ear density was lower on average by 16% (from 14-15% in cvs. Bombona and Vinjett to 20% in cv. Tybalt). Thousand grain weight of wheat cultivated in organic farming was also lower by 16% than that in conventional farming system. Thousand grain weight of Tybalt was reduced only by 9% while that of Parabola showed decrease by 22% (when in organic vs. conventional farming system).

Table 13: The yield of spring wheat and the components of yield in organic system

Cultivar	The yield of grain (t·ha-1)				Number of ears (pcs·m-2)				1000 grains weight (g)			
	2008	2009	2010	mean	2008	2009	2010	mean	2008	2009	2010	mean
Bombona	4.29	3.01	3.25	3.52	450	433	402	428	33.3	37.6	30.8	33.9
Bryza	4.13	2.63	2.71	3.16	454	363	357	389	34.0	38.1	32.1	34.7
Nawra	4.36	2.21	2.48	3.02	382	339	301	341	37.8	35.4	35.5	36.2
Parabola	4.31	2.69	3.21	3.40	385	341	277	334	39.3	41.2	40.9	40.5
Raweta	4.65	2.84	3.04	3.51	417	383	378	393	36.9	36.2	34.0	35.7
Tybalt	6.16	2.43	3.60	4.06	380	348	307	345	43.2	34.2	40.6	39.3
Vinjett	4.52	3.01	3.25	3.59	505	426	412	448	33.9	34.9	31.3	33.4
Zadra	4.81	3.05	3.20	3.69	422	392	361	392	32.8	35.6	31.6	33.3
ura	4.86	3.31	3.40	3.86	385	394	360	380	38.8	43.3	37.7	36.9
mean	4.68	2.80	3.13	3.54	420	380	351	383	36.7	37.4	34.9	36.0
HSD(α=0.05)	0.21	0.24	0.29	0.43	43	61	63	54	1.3	1.7	0.9	1.3

Table 14: The yield of spring wheat and the components of yield in conventional system

Cultivar	The yield of grain (t·ha-1)				Number of ears (pcs·m-2)				1000 grains weight (g)			
	2008	2009	2010	mean	2008	2009	2010	mean	2008	2009	2010	mean
Bombona	6.97	4.52	4.37	5.29	504	518	471	498	46.4	37.5	40.5	41.5
Parabola	7.40	4.07	5.06	5.51	370	447	398	405	56.9	48.2	51.1	52.1
Tybalt	7.75	4.37	5.67	5.93	512	430	421	455	48.9	37.3	43.5	43.2
Vinjett	6.71	4.34	4.77	5.27	550	507	513	524	44.4	33.8	37.2	38.5
mean	7.21	4.33	4.97	5.50	484	476	451	470	49.1	39.2	43.1	43.8
HSD(α=0.05)	0.52	0.18	0.53	0.64	64	54	61	37	1.6	1.0	1.9	2.1

Weed infestation of spring wheat in organic farming was on average 80 plants m^{-1} in 2008 and 2010 while in 2009 it was 115 plants m^{-2} (Table 15). Dry matter of weeds ranged from 7 g m^{-2} in 2008 (very successful wheat cropping) to 57 g m^{-2} in 2009 (worse stand density and uniformity due to unfavorable weather conditions). *Chenopodium album* and *Stellaria media* were the most common weed species throughout the years of the study. In 2010, weeds that are hard to control and more competitive to wheat were common *(Cirsium arvense, Anthemis arvensis* and *Galium aparine)*.

The competitiveness against weeds of spring wheat sown with the clovers and grasses was depended on morphological features of wheat varieties, plant density and biomass of undersown crop. Whether conditions were also a major factor affecting weed infestation as they affected the emergence rate of wheat and clover-grass mixture and its further development. A dense canopy of the undersown crop in 2008 resulted in a higher competitive abilities of spring wheat which had suppressed the growth of most weeds. On the other hand, a sparse stand of wheat in 2009, despite a good development of the undersown crop, resulted in weaker competitive abilities against weeds. Weed infestation in conventional farming system, where wheat was cultivated in pure sowing and herbicides were applied, was 4 times lower than in organic system (Table 15, 16). Also the dry matter of weeds was 3.5 times lower than that in organic farming. Cv. Vinjett showed statistically the lowest dry matter of weeds, while Parabola was the most weed-infested variety (table 16).

Table 15: Number and dry mater of weeds in spring wheat cultivated in organic system at dough stage

Cultivar (A)	Number of weeds (plants·m-2)				Dry matter of weeds (g·m-2)			
	Years (B)				Years (B)			
	2008	2009	2010	mean	2008	2009	2010	mean
Bombona	68.5	82.5	80.5	77.2	7.7	36.0	32.3	25.3
Bryza	106.5	142.5	92.5	113.8	7.5	78.1	57.8	47.8
Nawra	66.5	101.5	143.0	103.7	9.6	38.0	31.2	26.3
Parabola	77.5	84.5	79.0	80.3	7.6	79.0	57.8	48.1
Raweta	92.5	102.5	70.0	88.3	7.4	47.2	30.3	28.3
Tybalt	62.5	158.0	95.5	105.3	7.8	75.0	50.2	44.3
Vinjett	103.5	139.5	57.0	100.0	8.3	49.0	35.6	31.0

Zadra	58.5	123.5	85.5	89.2	2.5	61.0	37.3	33.6
ura	73.0	98.5	77.0	82.8	7.4	50.0	60.0	39.1
mean	78.8	114.8	86.7	93.4	7.3	57.0	43.6	36.0
HSD (=0.05) for:	A - 33.1, B - 14.6, interaction AB - 51.8				A – ns*, B - 14.9, interaction AB - ns			

[i] - * ns – non significant differences

Table 16: Number and dry mater of weeds in spring wheat cultivated in conventional system at dough stage

Cultivar (A)	Number of weeds (plants·m-2)				Dry matter of weeds (g·m-2)			
	Years (B)				Years (B)			
	2008	2009	2010	mean	2008	2009	2010	mean
Bombona	4.0	11.0	29.0	14.7	0.3	4.9	17.5	7.6
Parabola	12.5	35.0	20.5	22.7	2.1	34.9	26.5	21.2
Tybalt	13.0	32.5	47.5	31.0	1.3	4.4	20.9	8.9
Vinjett	8.5	35.5	17.0	20.3	1.1	5.4	7.0	4.5
Mean	9.5	28.5	28.5	22.2	1.2	12.4	18.0	10.5
HSD (α=0.05) for:	A – ns*, B – 18.1, interaction AB - ns				A – 16.2, B – 15.1, interaction AB - ns			

[i] - * ns – non significant differences

The differences in weed dry matter in spring wheat cultivated in organic farming were not statistically significant (Table 15). The competitive abilities of winter wheat varieties against weeds differ more than those of spring wheat varieties [2, 3, 30]. Bombona and Raweta were the tallest varieties and they also showed the lowest rate of weed infestation and dry weight of weeds (77 and 88 plants m^{-2}, 25 and 28 g m^{-2} respectively). Cvs. Bombona, Raweta and Vinjett were also characterized by a big density of plants and number of ears per unit area which could have resulted in a limited weed growth. Bryza and Tybalt were the most weed infested varieties (105 and 114 plants m^{-2}, 44 and 48 g m^{-2} respectively). Cv. Parabola also showed a high level of weed dry matter (Table 15). Cvs. Parabola, Tybalt and Nawra were characterized by the lowest number of ears per m^2 (Table 13).

Weed infestation is one of factors that can limit the yield of spring wheat. In 2008 the level of weed infestation was so low (less than 10 g m^{-2}) that it could not have significantly affected the yields of spring wheat (Figure 6). Cv. Tybalt, which was one of the least prone to stem and leaf fungal diseases (Table 17, 18), yielded the highest (6.16 t ha^{-1}), whereas the yield of the other varieties was similar (4.13-4.86 t ha^{-1}). In 2009, lower yields were found in those varieties which were more weed-infested than others (Tybalt, Bryza, Parabola – approximately 80 g of weed dry matter per 1 m^2) (Figure 7). In 2010, weed dry matter did not exceed 60 g m^{-2} and there were no significant differences in yield between varieties which may suggest that this level of weed infestation did not affect the yields (Figure 8). This is confirmed by the research results of Kapeluszny [31] in which the decrease of wheat yield occurred when weed infestation reached 96 plants per m^2 (62 g m^{-2} of dry matter) at dough stage. It is worth noting that cv. Nawra gave the lowest yields in 2009 and 2010 despite a small weed infestation, indicating that other factors could have affected the yields of this variety (Figure 7, 8).

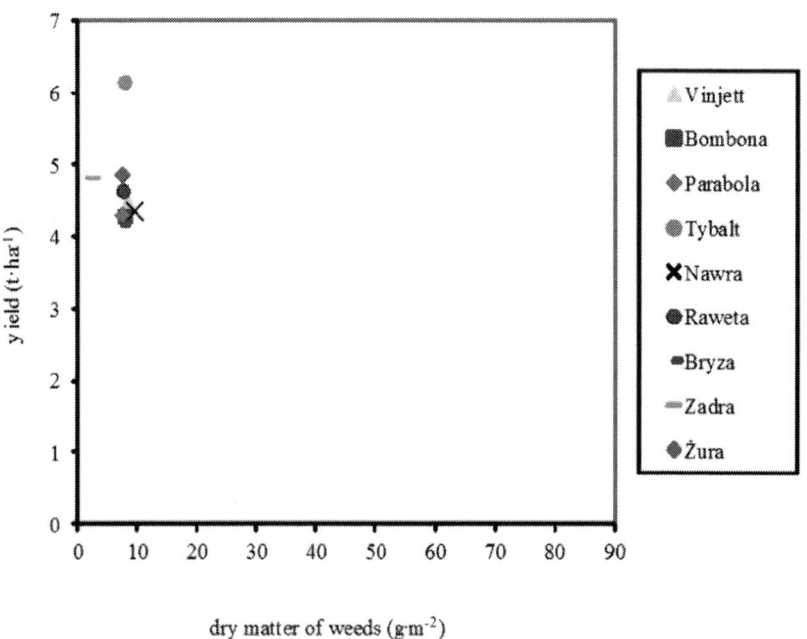

Figure 6: The dependence between weed dry matter and grain yield of spring wheat varieties cultivated in organic system in 2008.

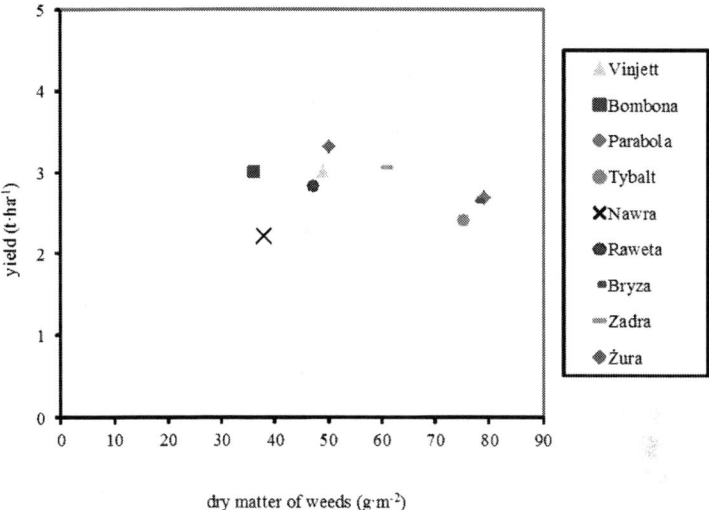

Figure 7: The dependence between weed dry matter and grain yield of spring wheat varieties cultivated in organic system in 2009.

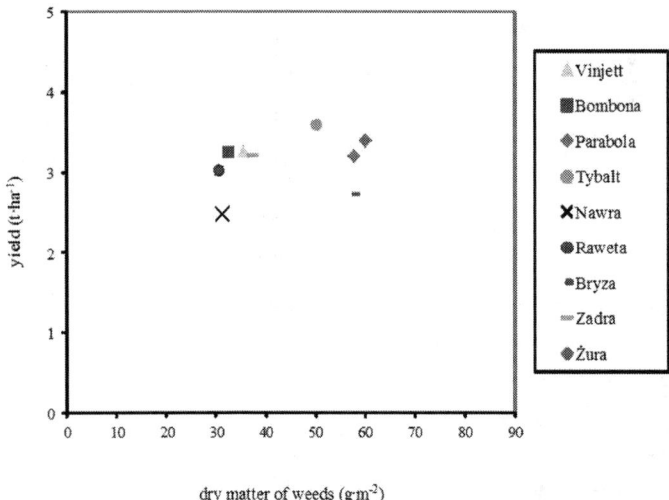

Figure 8: The dependence between weed dry matter and grain yield of spring wheat varieties cultivated in organic system in 2010.

Susceptibility to diseases is one of the factors that should be taken into account when evaluating the usefulness of cereal varieties in organic farming. The intensity of stem base diseases in spring wheat was low throughout the research years (Table 17). *Fusarium* spp. pathogens were the most common. According to the literature review, diseases caused by *Fusarium* are not more often in cereals grown in the organic system than in the conventional one (this also applies to mycotoxin content) [32, 33]. Cvs. Tybalt, Nawra and Żura were slightly more resistant to those diseases while Bombona and Parabola were more susceptible.

Table 17: The infestation index (%) of roots and stem base of spring wheat in organic system at milk-dough stage

Cultivar	Infection index (%)			Mean
	2008	2009	2010	
Bombona	5.7 ab*	8.6 b	24.8 c	13.0 b
Bryza	9.8 b	8.5 b	3.5 ab	7.3 ab
Nawra	7.7 ab	4.7 ab	1.5 a	4.6 a
Parabola	7.7 ab	3.2 ab	23.4 c	11.4 b
Raweta	3.2 a	3.8 ab	13.2 bc	6.7 a
Tybalt	3.8 a	2.4 a	6.7 ab	4.3 a
Vinjett	4.9 ab	4.1 ab	14.7 bc	7.9 ab
Zadra	7.4 ab	8.0 b	9.3 b	8.2 ab
Żura	7.4 ab	4.3 ab	4.0 ab	5.2 a
mean	6.4	5.3	11.2	7.6

[i] - * data marked with the same letters do not differ significantly at α=0.05

The damage of flag and underflag leaves by fungal pathogens was different among varieties and years, but in no case it was significant (Table 18). Cv. Tybalt showed the lowest (almost none) damage rate to its flag leaves at milk-dough stage, and its underflag leaves were just a little more damaged in 2009. On the other hand, Cv. Parabola was the most susceptible to fungal damage. With increasing fungal leaf diseases, the thousand grain weight decreased (Table 13). Incidence of fungal diseases in conventional farming system, due to the use of fungicides, was low. The thousand grain weight of conventionally cultivated wheat was higher by 14 to 31% than that of organic farming for cv. Parabola, and from 7 to 12% for cv. Tybalt.

Table 18: The area (%) of flag leaf and underflag leaf infested by pathogens in milk-dough stage in organic system

Cultivar	Flag leaf				Underflag leaf			
	2008	2009	2010	mean	2008	2009	2010	mean
Bombona	20.3	10.3	11.1	13.9	9.3	8.4	8.9	8.9
Bryza	14.8	11.8	11.7	12.8	8.0	3.6	4.6	5.4
Nawra	13.2	7.9	21.2	14.1	8.8	3.7	16.2	9.6
Parabola	25.2	13.8	27.3	22.1	17.4	3.3	7.9	11.2
Raweta	14.5	4.0	5.3	7.9	8.4	0.3	4.5	4.4
Tybalt	1.3	7.5	1.2	3.3	0	0.8	1.2	0.7
Vinjett	15.7	14.6	8.2	12.8	6.6	1.7	7.1	5.1
Zadra	16.5	12.1	12.1	13.6	7.3	2.6	5.9	5.3
ura	26.5	13.1	15.6	18.4	14.1	2.2	10.9	9.0
mean	16.4	10.7	12.6	12.7	8.9	3.1	7.5	6.5

Spring wheat was infested by different species of pathogens in different years. In 2008, *Puccinia recodita* was the most frequent disease. In that year, cvs. Żura and Zadra were also strongly infected with *Erysiphe graminis*. In 2009, there was no dominant pathogen species. Vinjett and Żura were infected with *Sepptoria* spp. and *Dreschlera tritici-repentis*, Bombona, Bryza and Zadra with *Puccinia recodita* and *Erysiphe graminis* while in Parabola all pathogen species occurred with similar intensity.*Puccinia recodita* and *Sepptoria* spp were the most common pathogens in 2010 in all varieties.

These small damage of flag and underflag leaves of spring wheat was due to favorable weather conditions. The sum of rainfall in June and July was generally lower than in the long-term average, while the average temperatures were significantly higher than in the long-term average (Table 12). Such pattern of weather conditions is not conductive to fungal diseases.

The analysis of correlation showed negative correlation between grain yield of spring wheat and weed infestation parameters at dough stage (Table 19). This correlation was weaker than that observed for winter wheat (Table 11). There was no significant correlation between grain yield and stem and leaf infestation by fungal pathogens (Table 19). Significant correlations between weed infestation and infection by fungal diseases were determined.

The cluster analysis based on grain yield, fungal infestation, weed abundance at dough stage allowed the division of varieties into 3 groups with similar characteristics (table 20).

The first cluster, represented by Bombona and Parabola, was characterized by the lowest yield, the highest infestation by fungal pathogens and an average level of weed abundance. In the second cluster, there were 6 varieties, characterized by high stand density and high number of ears per unit area, the lowest dry matter of weeds, an average infestation rate by fungal diseases and average yields. Cv. Tybalt was the only variety in the third cluster with the highest values for yield, tolerance of fungal diseases and weed infestation.

Table 19: Correlation coefficients (r) between grain yield and factors limiting the yield for all tested varieties grown in organic system (N=108)

Parameters	Grain yield (t·ha-1)	Density of ears (pcs·m-2)	Roots and stem base infestation index (%)	Leaves infestation index (%)	Number of weeds (pcs·m-2)	Dry matter of weeds (g·m-2)
Grain yield (t·ha-1)	x	0.379*	-0.108	-0.007	-0.385*	-0.558*
Density of ears (pcs·m-2)		x	-0.055	0.062	-0.058	-0.310*
Roots and stem base infestation index (%)			x	0.166	-0.253*	-0.018
Leaves infestation index (%)				x	-0.127	-0.267*
Number of weeds (pcs·m-2)					x	0.404*

Dry matter of weeds (g·m-2)						x

[i] - * significant at α=0.05

Table 20: The results of cluster analysis based on grain yield and factors limiting the yield for tested varieties of spring wheat cultivated in organic system

Cluster	Grain yield (t·ha-1)	Density of ears (pcs·m-2)	Roots and stem base infestation index (%)	Leaves infestation index (%)	Number of weeds (pcs·m-2)	Dry matter of weeds (g·m-2)	Cultivars
1	3.46	381.0	12.2	27.2	78.8	36.6	Bombona, Parabola
2	3.53	390.3	6.7	19.7	95.8	34.3	Bryza, Nawra, Raweta, Vinjett, Zadra, ura
3	4.06	345.0	4.3	4.0	105.3	44.3	Tybalt

The results obtained in the study does not give a clear assessment of suitability of the tested varieties to organic farming. As judged by the crop yields, cvs. Bryza and Nawra do not seem to be the best choice for the organic farming system (although they were as resistant to fungal diseases as the others). However, cv. Bryza showed lower competitive abilities against weeds that other varieties did. Also cv. Parabola, due to the higher infestation by weeds and diseases as well as because of small weight of grains (as compared to conventional system) is less suitable for organic farming system. On the other hand, cv. Tybalt should be preferred in the organic farming system due to its resistant to fungal leaf and stem diseases. Unfortunately, yields of cv. Tybalt were highly variable over the study years. It can also be assumed (although this was not investigated), that in the 3-year research period, availability of nitrogen and its uptake, besides weeds and disease infestation, had a strong influence on the yields of spring wheat varieties [12].

SUMMARY

- Among the nine compared varieties, cvs. Tybalt and Żura produced the highest yields, while cvs. Bryza and Nawra yielded the lowest.

- Crop yields were on average 34% lower in the organic system than in the conventional one. The differences were from 32% (cvs. Tybalt and Vinjett) to 38% (cv. Parabola). Lower yields in organic farming system were caused by lower ear density and thousand grain weight (both showing a reduction of approximately 16%).

- The weed infestation of spring wheat was strongly limited by undersown crops (clover-grass mixture). The dry matter of weeds averaged 36 g m^{-2} over the research years (from 7 to 57 g m^{-2}). The lowest dry matter of weeds was observed in Bombona, Raweta, Nawra and Vinjett which had a bigger density of canopy.

- In the 3-year period a minor incidence of stem base and leaf diseases was observed due to an appropriate crop rotation (stem base diseases) and favourable weather conditions in June and July (fungal leaf diseases). Cv. Parabola was more susceptible, while cv. Tybalt was more resistant to those diseases than the other varieties.

CONCLUSIONS

The yielding of cereals in organic system was lower than in conventional system. In case of winter wheat, the yield was on average 45% smaller in organic system than in conventional one. This low yield of winter wheat in organic system in years of research was mostly connected with unfavorable weather conditions in growing season 2009/2010. Winter wheat cultivars selected for organic farming should be characterized by high winter hardiness, because in this farming system is difficult to compensate for the effects of adverse weather conditions. Cereal canopy with a smaller plant density after winter is more susceptible to weed infestation. Earlier studies conducted on another set of winter wheat varieties (Roma, Kobra, Zyta, Sukces) showed a difference in the yields between organic and conventional systems by 19%. The difference in cereal yields in organic vs. conventional systems are due to lower density of ears, weed infestation, leaves diseases and nutrients deficiency [6, 9, 11, 12,14, 20].

Tested varieties were differed in their yielding and competitiveness against weeds as well as resistance to fungal diseases. Smuga variety gave the highest yield whereas Ostka Strzelecka and Spelt Schwabenkorn the lowest one. In case of spring wheat, 34% less grain yield on average was obtained in organic system than in conventional one. Among compared varieties of spring wheat cultivated in organic system, cvs. Tybalt and Żura, were characterized by the highest yield, while Bryza and Nawra had the smallest yield.

Analysis of correlation showed that grain yield of winter wheat and spring wheat was affected by number and dry matter of weeds. These relations were stronger for winter wheat varieties than for spring wheat varieties. Grain yield of winter wheat was also influenced by leaves infestation by fungal pathogens. There was no correlation between grain yield of winter and spring wheat in organic system and stem base diseases in the research period 2008-2010.

In the organic system weed infestation in cereals is generally greater than in the more intensive crop production systems where herbicides are used [34, 35]. However, the application of agricultural practices according to Good Agricultural Practices, proper crop rotation, delaying sowing time, increasing the amount of seed, maintaining good soil structure, with a high content of organic matter allows to keep the weeds at a level not causing a significant yield decrease [22, 36]. The results show high efficacy of weed control in spring wheat in the organic system by the interaction of 5-year crop rotation, successful undersown clover with grasses and dense canopy of wheat with large ability to compete with weeds. The effectiveness of mixed crops of cereals with legumes in reducing weed confirms the results of other authors [36, 37].

The competitive ability against weeds of winter and spring wheat varieties differed due to canopy parameters and morphological features. The level of weed infestation was influenced by parameters of wheat canopy: density and dry matter of wheat, as well as the height. Among modern varieties there are some with high competitive potential against weeds, for example cv. Smuga, which also has a greater yielding potential than old varieties. The competitive ability of spelt with weeds depends on the variety and habitat conditions [38].

Among spring wheat varieties the greatest suppressive ability against weeds were cvs. Bombona and Raweta.

In organic system stem base diseases were unimportant, contrary to leaves diseases. The dominant pathogens were: *Puccinia recondita* and *Septoria* spp. In case of spring wheat, greater susceptibility to fungal pathogens characterized cv. Parabola, while a smaller – cv. Tybalt. Among winter wheat varieties Kobra and Bogatka were the most resistant to fungal diseases leaves in organic farming conditions.

The old varieties of winter wheat: Ostka Kazimierska, Kujawianka Więcławicka, Wysokolitewka Sztywnosłoma had big potential to weed suppress, but were not very useful for cultivation in organic system, because they yielded 36% lower compare to modern varieties and were also strongly damaged by stem and leaf fungal diseases [20].

ACKNOWLEDGEMENTS

The studies have been supported by Ministry of Agriculture and Rural Development of Poland within the multi-annual program of Institute of Soil Science and Plant Cultivation-State Research Institute, task 3.2. Assessment of the directions and agricultural production systems and the possibilities of their implementation in the regions and farms

REFERENCES

1.   Kuś J. Jończyk K. Organic farming-state and prospects of its development. In: Agriculture XXI century-new aspects of farming. Institute of Animal Production (ed.), Kraków; 2010. p109-120.

2.   Eisele J.-A. Köpke U. Choice of cultivars in organic farming: New criteria for winter wheat ideotypes. Pflanzenbauwissenschaften 1997;1 19-24.

3.   Hoad S., Topp C., Davies K. Selection of cereals for weed suppression in organic agriculture: a method based on cultivar sensitivity to weed growth. Euphytica 2008;163(3) 355-366.

4.   Wolfe M. S., Baresel J. P., Desclaux D., Goldringer I., Hoad S., Kovacs G., Löschenberger F., Miedaner T., Østergård H., Lamberts

E.T. van Bueren. Developments in breeding cereals for organic agriculture. Euphytica 2008;163(3) 323-346.

5. Leibl M., Petr J. Varieties of winter wheat for ecological farming. In: Proceedings of the 13th International IFOAM Scientific Conference in Basel. Vdf. Hochschulverlag AG an der ETH Zurich; 2000 p243.

6. Kuś J., Mróz A., Jończyk K. Intensity of fungal diseases of selected varieties of winter wheat cultivated in the organic crop production systems. Journal of Research and Application in Agriculture Engineering 2006;51(2) 88-93.

7. Christensen S. Weed suppression ability of spring barley varieties. Weed Research 1995;35(4) 241-247.

8. Seavers G. P., Wright K. J. Crop canopy development and structure influence weed supression. Weed Research 1999;39 319-328.

9. Jończyk K. Response of selected winter wheat varieties for cultivation in different crop production systems. Pamiętnik Puławski 2002;130 339-346.

10. Bertholdsson N.-O. Early vigour and allelopathy – two useful traits for enhanced barley and wheat competiveness against weeds. Weed Research 2005;45(2) 94-102.

11. Baresel J. P., Zimmermann G., Reents H. J. Effects of genotype and environment on N uptake and N partition in organically grown winter wheat (*Triticum aestivum* L.) in Germany. Euphytica 2008;163 347-354.

12. Stalenga J. Evaluation of yielding, nutrient status and efficiency of nutrient uptake by selected modern and old winter wheat cultivars in organic crop production system. Journal of Research and Application in Agriculture Engineering 2009;54(4) 106-119.

13. Kolb L.N., Gallandt E.R. Weed management in organic cereals: advances and opportunities. Organic Agriculture 2012;2(1) 23-42.

14. Feledyn-Szewczyk B., Jończyk K., Berbeć A. The morphological features and canopy parameters as factors affecting the competition between winter wheat varieties and weeds. Journal of Plant Protection Research 2013;53(3) 203-209.

15. Cacak-Pietrzak G., Ceglińska A., Jończyk K. Technological value of selected varieties of winter wheat cultivated in different crop production systems. Pamiętnik Puławski 2003;133 17-25.

16. Lemerle D., Verbeek B., Orchard B. Ranking the ability of wheat varieties to compete with *Lolium rigidum*. Weed Research 2001;41 197-209.

17. Feledyn-Szewczyk B. Comparison of the competitiveness of modern and old winter wheat varieties in relations to weeds. Journal of Research and Application in Agriculture Engineering 2009;54(3) 60-67.

18. Didon U.M.E. Variation between barley cultivars in early response to weed competition. Journal of Agronomy and Crop Science 2002;188 176-184.

19. EPPO. Guidelines for the efficacy evaluation of plant protection products. Standards 1999;1 187-195.

20. Kuś J., Jończyk K., Stalenga J, Feledyn-Szewczyk B., Mróz A. Yields of the selected winter wheat varieties cultivated in organic and conventional crop production systems. Journal of Research and Application in Agriculture Engineering 2010;55(3) 219-223.

21. Seufert V., Ramankutty N., Foley J.A. Comparing the yields of organic and conventional agriculture. Nature 2012;485 229-232.

22. Tyburski J., Rychcik B. Weed infestation of winter wheat in conventional and organic farm on Elk Lake District. Pamiętnik Puławski 2007; 145 233-241.

23. Balyan R.S., Malik R.K., Panwar R.S., Singh S. Competitive ability of winter wheat cultivars with wild oat (*Avena ludoviciana*). Weed Science 1993;39 154-158.

24. Challaiah Burnside, O. C. and Wicks, G. A., Johnson, V. A. Competition between winter wheat (*Triticum aestivum*) cultivars and downy brome (*Bromus tectorum*). Weed Science 1986;34 689-693.

25. Satorre E.H., Snaydon R.W. A comparison of root and shoot competition between spring cereals and *Avena fatua* L. Weed Research 1992;32 45-55.

26. Wicks G.A., Ramsel R. E., Nordquist P.T., Smith, J.W., Challaiah R.E. Impact of wheat cultivars on establishment and suppression of summer annual weeds. Agronomy Journal 1986;78 59-62.

27. O'Donovan, J. T., Blackshaw, R. E., Harker, K. N., Clayton, G. W., McKenzie, R. Variable plant establishment contributes to differences in competitiveness with wild oat among wheat and

barley varieties. Canadian Journal of Plant Science 2005;85 771-776.

28. Worthington M., Reberg-Horton C.: Breeding cereal crops for enhanced weed suppression: Optimizing allelopathy and competitive ability. Journal of Chemical Ecology 2013;39(2) 213-231.

29. Kuś J., Jończyk K. The cultivation of cereals on organic farms. National Centre for Organic Agriculture (ed.), Radom 2003; 1-130.

30. Feledyn-Szewczyk, Berbeć A. K. Ranking the competitive ability against weeds of 13 spring wheat varieties cultivated in organic system in different regions of Poland. Journal of Research and Application in Agriculture Engineering 2013;58(3) 104-110.

31. Kapeluszny J. Development of yield structure and canopy of spring barley and spring wheat depending on the level of infestation. In: The causes and sources of infestation of cultivated fields: proceedings of the XVII National Conference, 28-29 June 1994, Olsztyn-Bęsia, Poland, ART (ed.) 1994; 95-100.

32. Sadowski Cz., Lenc L., Kuś J. Fusarium head blight and *Fusarium* spp. on grain of winter wheat, a mixture of cultivars and spelt grown in organic system. Journal of Research and Application in Agriculture Engineering 2010;55(4) 79-84.

33. Benbrook Ch.M. Breaking the mold – impacts of organic and conventional farming systems on mycotoxins in food and livestock feed. An Organic Center State of Center Review. The Organic Center 2005; 1-58.

34. Frieben B., Köpke U. Effect of farming systems on biodiversity. In: Isart J., Llerenea J.J. (eds.). Biodiversity and Land Use: The role of organic farming. Proceedings of the first ENOF Workshop, Bonn, 1995; 11-21.

35. Dąbkowska T., Stupnicka-Rodzynkiewicz E., Łabza T.: Weed infestation of cereals in organic, conventional and intensive farms in Małopolska Region. Pamiętnik Puławski 2007;145 5-16.

36. Feledyn-Szewczyk B. The effectiveness of weed regulation methods in spring wheat cultivated in integrated, conventional and organic crop production systems. Journal of Plant Protection Research 2012;52(4) 486-493.

37. Hauggaard-Nielsen H., Ambus P., Bellostas N., Boisen S., Brisson N., Corr-Hellou, Crozat Y., Dahlmann C., Dibet A., Fragstein P., Gooding M., Kasyanova E., Launay M., Monti M., Pristeri A., Jensen E.S. Intercropping of pea and barley for increased production, weed control, improved product quality and prevention of nitrogen-looses in European organic farming systems. Bibliotheca Fragmenta Agronomica 2006;11(III) 53-60.

38. Sulewska H., Koziara W., Panasiewicz K., Ptaszyńska G. Yielding of two spelt varieties depending on sowing date and sowing rate in central Wielkopolska conditions. Journal of Research and Application in Agriculture Engineering 2008;53(4) 85-91.

Chapter 5

Use of CLIMEX, Land use and Topography to Refine Areas Suitable for Date Palm Cultivation in Spain under Climate Change Scenarios

Farzin Shabani[1], Lalit Kumar[1], and Atefeh Esmaeili[2]

[1]Ecosystem Management, School of Environmental and Rural Science, University of New England, Armidale, NSW, 2351, Australia

[2]Soil Science Department, Faculty of Agricultural Engineering and Technology, University College of Agriculture and Natural Resources, University of Tehran, Tehran, Iran

ABSTRACT

In this study, CLIMEX modeling software was used to develop a model of the potential distribution of P. dactylifera under current and various future climate scenarios for Spain. CLIMEX parameters were adjusted depending on satisfactory agreement between the potential and known distribution of P. dactylifera in northern African countries, Iraq, Saudi Arabia, Oman and Iran. The potential date palm distribution was modeled under current and future climate scenarios using one emission scenario (A2) with two different Global Climate Models (GCMs): CSIRO-Mk3.0 (CS) and MIROC-H (MR). The CLIMEX outputs were then refined by land use types and areas less than 10 slope, since sloping areas impose problems in hydraulic conductivity and root development. The refined results indicated that large areas in Spain are projected to become climatically more suitable for date palm growth by 2100. However, the results from the CS and MR GCMs show some disagreements. The refined MR GCM projected that approximately 22.86 million hectares in Spain may become suitable for date palm growth, while the CS GCM showed approximately 18.72 million hectares by 2100. The refined results showed that only about 65% of CLIMEX results are suitable for date palm cultivations while the rest of the areas are unsuitable due to the unsuitability of land uses and slope. Our results indicated that cold and wet stresses will play a significant role in date palm distribution in some central and northern regions of Spain by 2100.

INTRODUCTION

Global climate change resulting from changes in air and sea surface temperatures [1]; precipitation patterns, ocean level alterations [2] and ocean salinity [3, 4] can affect agricultural productivity [5-7]. A review of literature reveals there have been many studies on the effect of climate change on ecological plant processes, such as plant growth [8], crop yield [9] and plant community interactions [3,10-14]. While McDermott [15] documented a greater than 55% crop failure in agricultural regions in India as a consequence of climate change, a comparable study reports that Sudan experienced a 50% crop failure in 2009 as a consequence of small changes in the quantity and pattern of precipitation, resulting from climate change [16]. Further, the potential effect on maize production of climate change has been reported [17] with indications of a 10% reduction by 2055, resulting in a loss of $2 billion per year in Africa and Latin America [17]. Shabani et al. [18] have reported on the effect of altered climate on global date palm production, in terms of long-term broad-scale shifts in the areas conducive to date palm cultivation. The total annual income from date palms in the Middle Eastern countries decreased from 1990 to 2000, due to water shortage and plant diseases resulting from climate change [19]; while an earlier example indicates that climate change caused a $438 million loss in wheat, a $116 million loss in grapes and a $67 million loss in sugar production in Australia and North America [14]. Substantial challenges for development and food security are posed by such losses in production and necessitate modeling the effects of climate change on agricultural production, especially crops important economically, to study the probable impact on agricultural output and distribution.

Date palm, which includes approximately 400 different species within the Arecaceae family, is a crop of economic importance in Middle Eastern countries, including Saudi Arabia, Egypt, Iran, Iraq, as well as some European countries, such as Turkey and Spain [20-24]. Date palm, being suited to a Mediterranean climate with salty and alkaline soils, has been grown in Spain for over 150 years and has become one of the main income sources for local farmers and the Spanish government [25]. Elche, located on the east coast of Spain, is one of Europe's largest date palm plantations [26,27]. A statistical report indicates that date palm production in Spain reached 3,732 metric tons in 2003 [28]. Climatic parameters, plant diseases resulting from climate change and water shortage are factors that will impact on future date palm yields. Thus identifying how different regions of Spain may be affected by climate change is vital in terms of the future of date palm production in Spain.

Global climate models (GCMs) are important analytical tools in such disciplines as biogeography, evolution, ecology, conservation and invasive species management [29] and extensive use has been made of GCMs in biogeography, conservation biology and environmental management studies since 1996 [3,8,18,30-34]. The major function of climatic models is to illustrate plant sensitivity to climate change, through ascertaining relationships between plant and atmospheric sciences [3]. Such models use the current climate range to identify climate tolerance limits of particular species. Then, by modeling climate change scenarios, the alteration of the areas suitable for growth of vulnerable species can be projected into the future [35]. Such projections are valuable for the development of agricultural organizations' long-term management strategies for sustainable production of economically important crops such as date palm (Phoenix dactylifera L.) under predicted future climate conditions [18].

There have been many models used to address the climate change factors, including bioclimatic models, ecological niche models (ENMs) [36], species distribution models (SDMs) [36-38] and CLIMEX [39]. CLIMEX is a computer-based software, designed for matching climates in ecology and has been widely used in estimating potential geographical distribution of different species in the future [11,40-44]. With CLIMEX, users can identify areas where selected species could establish and maintain or develop, based on predicted climate alterations. Through comparisons with other correlative modeling methods [45], CLIMEX has been evaluated as the most successful climate modeling software, for describing species' responses to climate, due to its geographic range and powerful phonological observations [18,40,46,47].

Criticism of CLIMEX indicates that it does not include biotic interactions and dispersal in the modeling process [10,13]. Thus CLIMEX modeling output is based on limited climatic variables, such as overall maximum and minimum monthly temperatures (Tmax and Tmin), overall monthly precipitation levels (Ptotal) and the relative humidity taken at 09:00 h (RH09:00) and 15:00 h (RH15:00) [46,48]. This means that CLIMEX outputs are only based on climate predictions and the projected areas may still be unsuitable due to variables such as soil types, high slopes and urban areas. Thus, refining date palm suitability regions modeled with CLIMEX by investigating slopes and land use would result in a more accurate identification of the areas available for future date palm cultivation. Such refining can be undertaken using geographic information systems (GIS), remote sensing software (RS), species distribution models (SDMs) and ecological niche models (ENMs) [18,30,49-53]. In this study, it has been proposed that the distribution of P. dactylifera may effectively change as a result of climate alteration. With this possible outcome in mind, this research made use of the CLIMEX software package to determine the potential distribution of P. dactylifera under current and future climate and the location of the suitable slopes and land uses to assess the sensitivity of its distribution to climate change, and to evaluate the concurrent implications for cultivation of date palms for 2030, 2050, 2070 and 2100 in Spain.

METHODOLOGY

Distribution of Date Palms (*P. Dactylifera*)

Landsat images, with 30 m image resolution [54], the Global Biodiversity Information Facility (GBIF) [55], Missouri Botanical Gardens' database [56] and other date palm literatures in CAB Abstracts databases [57] were used to collect data on P. dactylifera distribution and this information was supplemented by other date palm literature [22,24,58-70]. The GBIF database for Northern African countries, Saudi Arabia, Oman, Iraq and Iran had 314 records for P. dactylifera, but 64 records did not have geographical coordinates and were therefore removed, leaving 250 records. Some of these records were duplicates and were also deleted. In the end, 147 records from the GBIF database, 33 records from Missouri Botanical Gardens' database and 43 records obtained from literature research were used to validate parameters in CLIMEX.

CLIMEX Software, Climate Data, Climate Models and Climate Scenarios

As one of the most comprehensive inferential modeling software packages, CLIMEX allows users to project the climatic impact on potential distribution of different species, based primarily on their current distribution [71]. The program makes use of known associations among climate, species distribution and their biological responses to project their future distributions [72]. Experimental observations of species' growth response to temperature and soil moisture can be taken into account in CLIMEX [8,30,73,74]. The Ecoclimatic Index (EI) in CLIMEX is an average yearly index of climatic suitability based on regular weekly calculations of growth and stress indices. EI values range from 0 to 100, and a species can be established in an area if EI>0. EI values ranging from 0 to 10 indicate marginal potential habitats, while EI values >20 are highly favorable predictors [40]. Thus, an area with an EI>20 represents a potential area of high suitability from a climatic perspective, for introducing a particular species.

A variety of climatic parameter averages recorded between 1950 and 2000 are include in the CliMond 10′ gridded climate records and these were used to model the current distribution of date palm [46]. The meteorological database's climatic parameters include overall maximum and minimum monthly temperature (Tmax and Tmin), overall monthly precipitation (Ptotal) and relative humidity as recorded at 09:00 h (RH09:00) and 15:00 h (RH15:00). These parameters were also used to predict climates for the mid- to late 21st century. The potential distribution of date palms under future climate was modeled using two Global Climate Models (GCMs), namely, CSIRO-Mk3.0 and MIROC-H (Center for Climate Research, Japan), with the A2 SRES scenario. These two GCMs were part of the CliMond dataset and were selected from 23 GCMs based on criteria such as the availability of required variables (including temperature, precipitation, sea level pressure and humidity for CLIMEX), small horizontal grid spacing in both GCMs and better representation of observed climate at local scales, compared with other GCMs [2,75,76].

The A2 SRES scenario was chosen as a possible climate scenario principally because of its inclusion of demographic and financial factors, as well as incorporating technological forces driving GHG emissions. The A2 scenario assumptions take into account factors of independent and self-reliant nations, such as population increases and regional economic development [77]. Future assumptions inherent in A2 SRES include high population growth, coupled with slow economic growth and extensive technological change. The A2 scenario assumes less extreme range of GHG emissions than scenarios such as A1F1, A1B, B2, A1T and B1 [77, 78].

Model Framing

The global distribution of date palm and the suitable areas for date palm cultivation at a global scale have been recently reported by Shabani et al. [18]. The reported CLIMEX parameters were used for Northern African countries, Saudi Arabia, Oman, Iraq, and Iran as the validation areas. Here, temperature and moisture response parameter values were obtained from the date palm physiology and growth model literature [16,22,35,58,67,68,79,80] and transformed into CLIMEX-compatible Temperature and Moisture Index parameters and cold, heat, dry and wet stress threshold values. The final values used for climatic parameters and growth indices in CLIMEX were 0.9 for the wet stress threshold (SMWS), 0.022 week–1 for the accumulation rate of wet stress (HWS), 4°C for the cold stress temperature threshold (TTCS), –0.01 week–1 for the frost stress accumulation rate (THCS), 46°C for the heat stress parameter (TTHS), 0.9 week–1 for the heat stress accumulation rate (THHS), 14°C for the limiting low temperature (DV0), 20°C for the lower temperature (DV1), 39°C for upper optimal temperatures (DV2), 46°C for the limiting high temperature (DV3), 0.007 for the limiting low moisture threshold (SM0), 0.013 for the optimal soil moisture (SM1), 0.81 for the upper optimal soil moisture thresholds (SM2) and 0.9 for the limiting high soil moisture (SM3) [24,25,55,58,62,65,70]. The data sets were output from CLIMEX and imported into ArcGIS (GIS) software for further processing and mapping.

Refining the CLIMEX Outputs by Suitable Land Uses and Slopes

All vector data representing unsuitable land uses in Spain were taken from European Environment Agency Datasets [81]. The digital elevation model (DEM) of Spain on a regular 30 m grid from ASTER [82] (NASA, 2012) database were used to generate a slope surface, which was then reclassified into the two slope categories, mainly because overlaying the satellite images onto the date palm sites showed that 90% of the locations were on slopes less than 10 degree slopes (suitable <10 degrees and unsuitable >10 degree). Slope classes in raster format were converted to polygon shape files and queries were undertaken using attribute and location of unsuitable land uses including construction sites, coniferous forest, mineral extraction sites, rice fields, salt marshes, road and rail networks and associated land, green urban areas and bare rocks to extract the areas having suitable slopes and suitable land uses. The CLIMEX outputs for current time were overlaid on the location of suitable slopes and suitable land uses for the whole country and all those locations that satisfied the condition of EI>20 were selected.

RESULTS

Model Validation under Current Climate

The distribution of P. dactylifera taken from different databases and an appropriate match between the Ecoclimatic Index (EI) from the CLIMEX model for Morocco, Algeria, Libya, Egypt, Saudi Arabia, Oman, Iraq and Iran is illustrated in Figure 1. The modeled results showed that southern Iran, Oman, Saudi Arabia, northern Egypt, northern Libya, Algeria and western Morocco have suitable climatic condition for P. dactylifera. While large parts of Libya and Algeria are modeled to have suitable climatic conditions for P. dactylifera in its current known distribution, inadequate distribution data were available from these regions. This could be due to a shortage of reporting from these areas as a result of lack of human distribution, or biotic factors such as competition or lack of dispersal opportunities could prevent this species from occurring in these areas [31,83-85]. Nearly 89% of the occurrence records fell within the suitable categories, confirming that the values selected for the various parameters in CLIMEX were optimum (Figure 1).

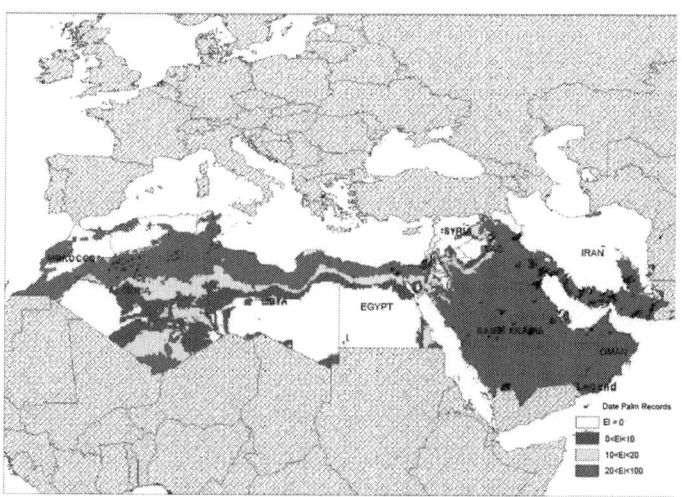

Figure 1: Current and potential distribution of P. dactylifera in a validation region based on the EI index.

Future Climate

Figures 2 and 3 are illustrations of the Ecoclimatic Index for P. dactylifera, using the two global climate change models (CS and MR) and the A2 emission scenario, and show the projected distribution for the present, 2030, 2050, 2070 and 2100. The CS modeled results show that large areas of southern, western and central regions of Spain are predicted to become climatically more suitable for date palm cultivation and production between 2030 and 2100. The same model indicates that areas of northeast Spain may become suitable for P. dactyliferas' cultivation between 2050 and 2100. The CS GCM predicts that northern and western Spain will become climatically unsuitable for date palm cultivation by 2100. This model shows roughly 11.5, 14, 19.5 and 28 million hectares will be conducive for date palm growth by 2030, 2050, 2070 and 2100 respectively (Table 1 and Figure 2). The CS model demonstrates that the areas located in 37.5 to 40N and 1 to 5W will become highly suitable for date palm cultivation from 2070 to 2100 while in this region the cold stress will remain and impose limitation for date growth until 2070 (Figures 2 and 6). Using the MR GCM, the model projects large areas in central and southwestern Spain will be climatically suitable for date palm growth by 2100. Additionally, the MR GCM results indicate that some areas in northern Spain may have an opportunity for date palm cultivation. As the climate condition is projected to be suitable by 2100. From the MR GCM, it can be seen that northwestern Spain will be climatically unsuitable, similar to the current situation. This model shows that areas from 37.5 to 40 N and 1 to 5 W and 40 N to 43 N and 4 W to 7 W will become highly conducive for date palm cultivation by 2100 (Figure 3). The MR GCM shows 12.2, 16, 24 and 33.8 million ha of Spain may have suitable climatic conditions for date palm cultivation by 2030, 2050, 2070 and 2100, respectively.

Generally, both CSIRO-Mk3.0 and MIROC-H GCMs project similar trends for Spain. However, some differences can be seen in the projections. These differing results are due to the different predictions of future climate by the two GCMs [86,87] (Figure 2 and 3 and Table 1).

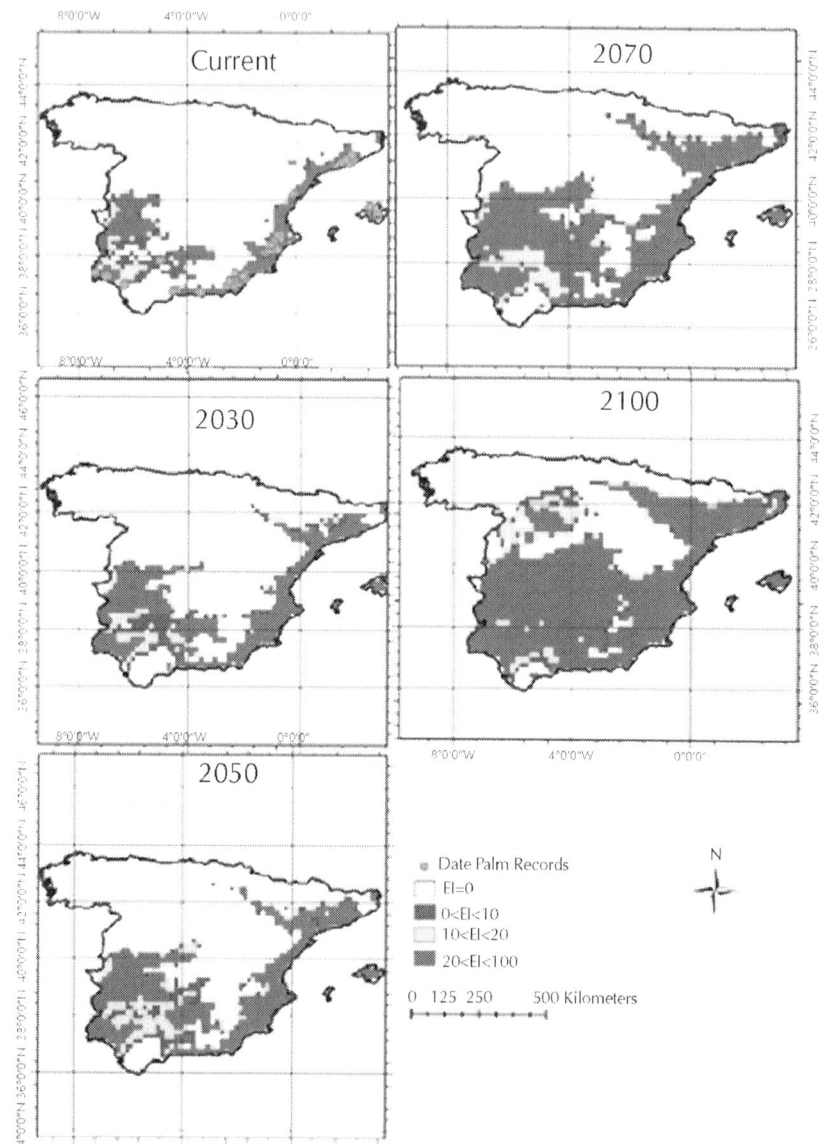

Figure 2: The climate (EI) for *P. dactylifera* at present and projected using CLIMEX under the CSIRO-Mk3.0 GCM running the SRES A2 scenario and for 2030, 2050, 2070 and 2100 for Spain.

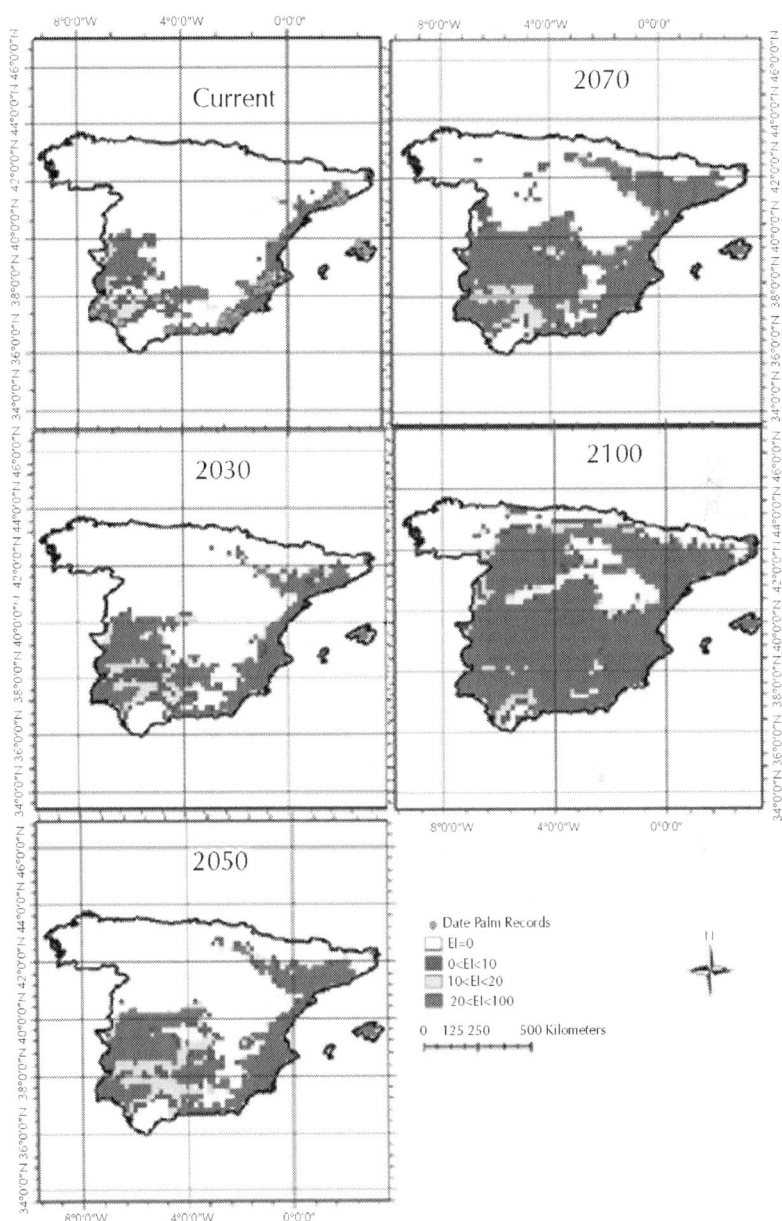

Figure 3: The climate (EI) for *P. dactylifera* at present and projected using CLI-MEX under the MIROC-H GCM running the SRES A2 scenario and for 2030, 2050, 2070 and 2100 for Spain.

Future Climate + Refined Results

The refined CS outputs by areas less than 10°slope and the location of suitable land use for date palm cultivation shows that only about 7.38, 8.80, 12.81 and 18.72 million ha will be highly suitable for date plantation by 2030, 2050, 2070 and 2100 in Spain. The refined MR results show that about 7.87, 10.34, 15.14 and 22.86 million ha will be highly conducive for this species growth (Table 1).

Table 1: Table 1- Results of CS and MR GCMs and the refined outputs by the location of unsuitable land uses and area less than 10 slopes for date palm cultivation for 2030, 2050, 2070 and 2100

Years	Projected by CS(GCM)	Projected by MR(GCM)	CS+Slope+land use(Milion ha)	MR + Slope + land use(Milion ha)
2030	11.55	12.2	7.38	7.87
2050	13.77	15.88	8.80	10.34
2070	19.49	24.12	12.81	15.14
2100	18.12	33.80	18.72	22.86

DISCUSSION

Species distribution modeling was used as a tool with the aim of illustrating new conducive regions for date palm cultivation in Spain in the future. The results showed that, under current climate, large parts of eastern and western Spain are highly conducive to date palm cultivation (Figure 2), as there are no cold and wet stresses in those regions under the current climate (Figure 6).

There was some agreement in the projections of suitable areas for date palm cultivation between the CS and MR GCMs for Spain. Both MR and CS GCMs project that large parts of Spain will become more climatically suitable for date palm cultivation towards 2100 as a result of significant reduction in areas where currently cold and wet stresses are the major factors restraining date palm distribution (Figures 2,3 and 6). For example, both models project that the areas with the highest potential for the establishment of date palm by 2100 are central Spain because cold stress may shift northwards (Figure 6). Additionally, both models projected that this unsuitability may be limited to the northern and northwestern regions of Spain due to significant increases in wet stress (Figure 6).

A comparison of the MR and CS models shows some projection differences. While the MR GCM projected that most regions in central Spain could become climatically suitable for date palm growth, the CS GCM projections showed similar results but with less extreme change and effects of change, especially from 2070 to 2100. It should be noted that some differences in the results of CS and MR GCMs (Figure 2 and 3) are due to differences in future greenhouse emission patterns. In this context, the MR model predicts that temperatures will increase by approximately 4.31°C, while the CS model predicts an increase of only 2.11°C by 2100. While the CS model predicts a 14% decrease in future mean annual rainfall, the MR model predicts a mere 1% reduction [2,86,87].

Climate change modeling results indicate the probable changes in the potential distribution of P. dactylifera in the future. As climate changes, certain areas where P. dactylifera currently occurs may become climatically unsuitable, and the economies of those areas will consequently be affected. The results of this study thus have broad implications for future date palm production in Spain for local farmers and the Spanish government, in terms of the predicted increasing possibilities for cultivating date palm. Predictions suggest there will be more areas available that are unaffected by wet or cold stresses, and thus be more conducive for date palm cultivation. In other words, Spain will be able to cultivate these crops to a larger extent in the future compared with the present.

The necessity of refining the CLIMEX results using the slope factor was due to the observation of high correlation between species distribution and abundance as functions of topography which directly effects moisture and fertility gradients [88,89]. Also, most of the studies on distribution of tree species have shown that topography is one of the most important factors affecting yields by causing problems in air permeability, hydraulic conductivity and root developments [90]. Furthermore, many studies have concluded that transportability of the sediment as a consequence of slope creates diverse microsite depending upon the soil type, geology and latitude [89, 91, 92].

A comparison between the mean CS GCM output and the mean refined results using slope and land use shows that only about 65% of the CS outputs will be suitable for date palm cultivation because of unsuitability in slopes and land uses. The slope map indicates that 34 million ha of Spain has less than 10 slopes and the Spain land use map showed that 2 million ha has highly unsuitable land use for date palm cultivation. For instance, wetland in Spain covers 1.1 million ha and this cannot be used as suitable land for date palm cultivation. From the MR results, refined using the suitable slope and land use, it can be seen that only 65% of the area which was projected to be climatically suitable may be conducive for date palm cultivation while the rest will be unsuitable due to unsuitable slopes and land uses.

A comparison of the results generated by the two models and then refined by slope and land use showed an average difference of about 15%. In the CLIMEX modeled results, the areas that are commonly projected by both GCMs was about 67%. Once the two modeled outputs were refined utilizing land use and slope data, results showed that 83% of areas projected by the two models overlapped for 2030, 2050, 2070 and 2100 (Figure 4 and 5).

It should be noted that there is a need to take into account additional non-climatic factors such as water access to the CLIMEX modeling output. For example, date palm can definitely cope with drought since it has evolved some strategies including high level of water absorption and widespread root system to protect itself in the face of stressors. However, in some areas date palm could only be cultivated (for fruit production) because it is irrigated. Thus, when there is no rainfall or irrigation, fruit production is greatly diminished. Therefore, water availability is an important non-climatic factor and could be a suitable refinement tool to refine the CLIMEX modeling results.

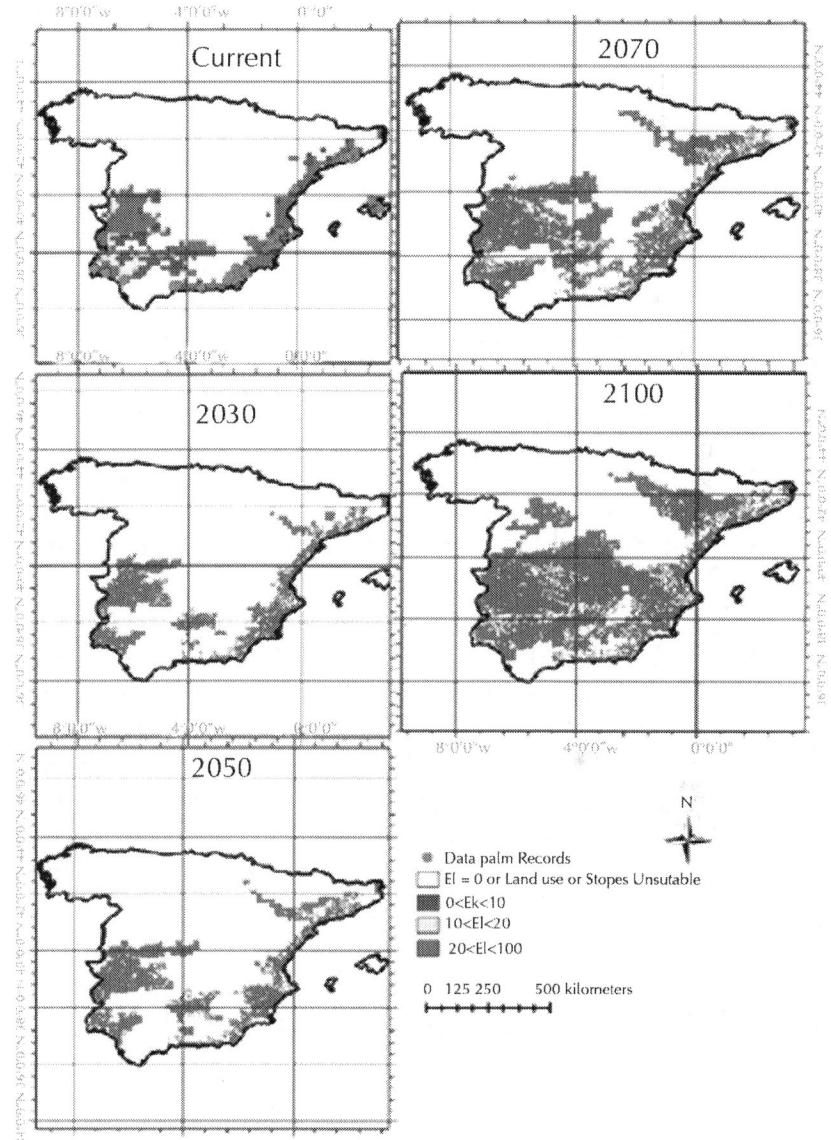

Figure 4: The refined suitable areas for *P. dactylifera* under the CSIRO-Mk3.0 GCM running the SRES A2 scenario for 2030, 2050, 2070 and 2100 by the location of suitable land use and less than 10 slopes.

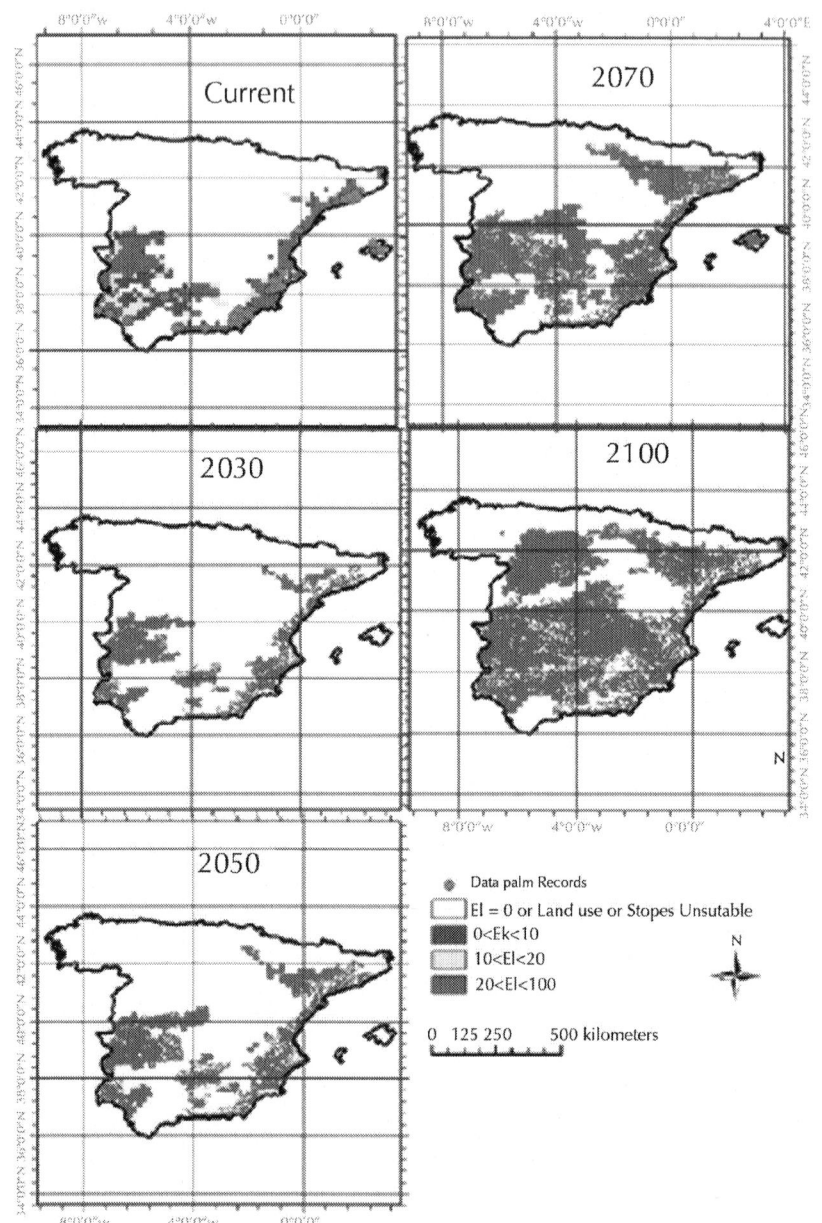

Figure 5: The refined suitable areas for *P. dactylifera* under the MIROC-H GCM running the SRES A2 scenario for 2030, 2050, 2070 and 2100 by the location of suitable land use and less than 10 slopes.

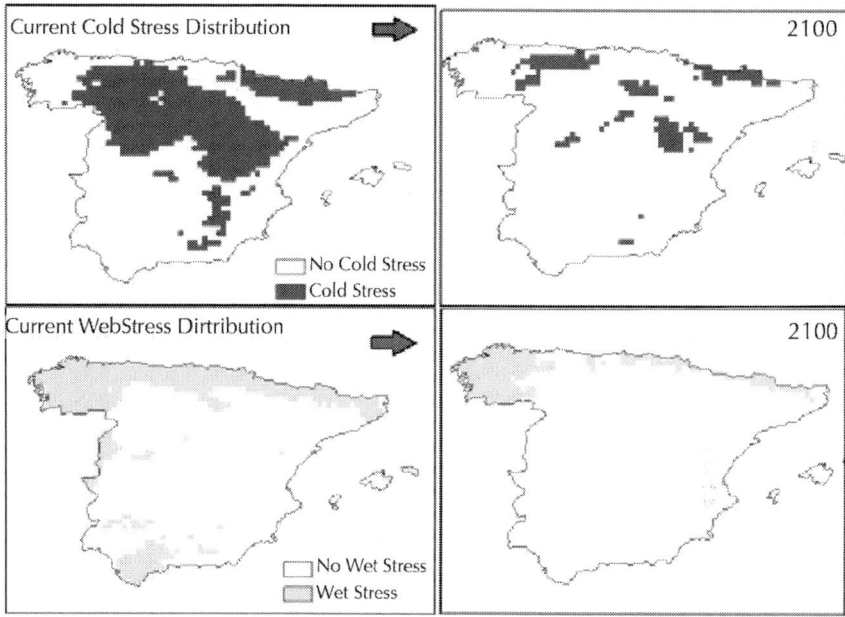

Figure 6: Location of cold and wet stresses in Spain for date palm cultivation for the present time and 2100.

CONCLUSIONS

Such modeling allows us an insight into potential impacts of climate change on agricultural production and enables us to take appropriate steps to minimize negative impacts, as well as to take advantage of some scenarios that may arise in the future. Distribution maps such as those used to illustrate this study can assist in the formulation of production methods for cost-effective agricultural crops of the future. Valuable information on current and future changes of areas suitable for date palm cultivation can be deduced from maps such as those shown in Figure 4 and 5 and may encourage plantation owners and managers to consider the future implications of present day management decisions. The methods utilized in this study have equal application for other agricultural crops. Major factors that contributed to the success of this study were: (a) all variables needed for CLIMEX, including temperature, precipitation, sea level pressure and humidity, were readily available; (b) well-documented historical data on date palm distribution was incorporated; and (c) the popular GCMs chosen for this study were on account of (I) their small horizontal grid spacing and (II) their better representation of observed climate at the local scale, compared with GCMs not used in this study. The limitations of this study should be noted. For example, CLIMEX results are based on the response of a species to climate, and non-climatic parameters such as water availability, biotic interactions, soil type and competition between species have not been taken into account. Thus areas shown as being suitable are only based on future climate, suitable slopes and land uses. Large parts of these areas may be unsuitable due to the unsuitability of soil types. Future research could consider these factors in addition to CLIMEX-based modeling.

REFERENCES

1.   Jeffrey S, Harold A (1999) Does global change increase the success of biological invaders? Trends in Ecology and Evolution 14: 135-139.

2.   Hennessy K, Colman R (2007) Global Climate Change Projections. Available.

3.   Adams M, Rosenzweig C, Peart M, Ritchie T, McCarl A, et al. (1990) Global climate change and US agriculture. Nature 345: 219-224.

4.   Mas-Coma S, Valero M, Bargues M (2009) Climate change effects on trematodiases, with emphasis on zoonotic fascioliasis and schistosomiasis. Veterinary Parasitology 163: 264-280.

5.   Bradley BA, Wilcove DS, Oppenheimer M (2010) Climate change increases risk of plant invasion in the Eastern United States. Biological Invasions 12: 1855-1872.

6.   Fordham DA, Akçakaya HR, Araújo M, Brook BW (2012) Modelling range shifts for invasive vertebrates in response to climate change. Wildlife Conser Chang Clim: 86.

7.   Luedeling E, Gebauer J, Buerkert A (2009) Climate change effects on winter chill for tree crops with chilling requirements on the Arabian Peninsula. Clim Chang 96: 219-237.

8. Pearson RG, Dawson TP (2003) Predicting the impacts of climate change on the distribution of species: Are bioclimate envelope models useful? Global Ecology and Biogeography 12: 361-371.

9. Scherm H, Sutherst RW, Harrington R, Ingram JSI (2000) Global networking for assessment of impacts of global change on plant pests. Environmental Pollution 108: 333-341.

10. Araújo MB, Luoto M (2007) The importance of biotic interactions for modelling species distributions under climate change. Global Eco Biogeogra 16: 743-753.

11. Kriticos DJ, Sutherst RW, Brown JR, Adkins SW, Maywald GF (2003) Climate change and the potential distribution of an invasive alien plant: Acacia nilotica ssp indica in Australia. J Appl Eco 40: 111-124.

12. Visser ME, Both C (2005) Shifts in phenology due to global climate change: The need for a yardstick. Proceedings of the Royal Society B: Biological Sciences 272: 2561-2569.

13. Barney JN, DiTomaso JM (2011) Global Climate Niche Estimates for Bioenergy Crops and Invasive Species of Agronomic Origin: Potential Problems and Opportunities. PLoS ONE 6: e17222

14. Chakraborty S, Murray GM, Magarey PA, Yonow T, O'Brien RG, et al. (1998) Potential impact of climate change on plant diseases of economic significance to Australia. Austral Plant Path 27: 15-35.

15. McDermott M (2009) Climate Change-Induced Drought Causing Crop Failure, Livestock Problems in Indian Himalayas, India.

16. Elshibli S, Elshibli E, Korpelainen H (2009) Date Palm (Phoenix dactylifera L.) Plants under Water Stress: Maximisation of Photosynthetic CO_2 Supply Function and Ecotypespecific Response. "Biophysical and Socio-economic Frame Conditions for the Sustainable Management of Natural Resources" Tropentag, Hamburg.

17. Jones G, Thornton K (2003) The potential impacts of climate change on maize production in Africa and Latin America in 2055. Global Environ Chang 13: 51-59.

18. Shabani F, Kumar L, Taylor S (2012) Climate Change Impacts on the Future Distribution of Date Palms: A Modeling Exercise Using CLIMEX. PLoS ONE 7: e48021.

19. Botes A, Zaid A (2002) Date palm cultivation.

20. Ferry M, Gomez S (2002)The Red Palm Weevil in the Mediterranean Area. J Int Palm Soc 46.

21. Gómez-Vidal S, Lopez-Llorca LV, Jansson HB, Salinas J (2006) Endophytic colonization of date palm (Phoenix dactylifera L.) leaves by entomopathogenic fungi. Micron 37: 624-632.

22. Tengberg M (2011) Beginnings and early history of date palm garden cultivation in the Middle East. Journal of Arid Environments 5: 1-9.

23. Ahmed M, Bouna Z, Lemine F, Djeh T, Mokhtar T and et al. (2011) Use of multivariate analysis to assess phenotypic diversity of date palm (Phoenix dactylifera L.) cultivars. Scientia Horticulturae 127: 367-371.

24. Bokhary H (2010) Seed-borne fungi of date-palm, Phoenix dactylifera L. from Saudi Arabia. Saudi Journal of Biological Sciences 17: 327-329.

25. Chao C, Krueger R (2007) The Date Palm (Phoenix dactylifera L.): Overview of Biology, Uses, and Cultivation. Journal of Hortscience 42: 1077-1083.

26. Veramendi J, Navarro L (1996) Influence of physical conditions of nutrient medium and sucrose on somatic embryogenesis of date palm. Plant Cell, Tissue Organ Cul 45: 159-164.

27. Robinson M, Brown B, Williams C (2012) The date palm in southern Nevada. Nevada: The University of Nevada. 23

28. Erskine W, Moustafa A, Ahmad E, Lashine A, Badawi T and et al. Date Palm in the GCC countries of the Arabian Peninsula, Egypt: International Center for Agricultural Research in the Dry Areas.

29. Anderson P, Lew D, Peterson A (2003) Evaluating predictive models of species' distributions: criteria for selecting optimal models. Ecological Modelling 162: 211-232.

30. Taylor S, Kumar L, Reid N, Kriticos DJ (2012) Climate Change and the Potential Distribution of an Invasive Shrub, Lantana camara.L. PLoS ONE 7: e35565.

31. Brooker RW, Travis JMJ, Clark EJ, Dytham C (2007) Modelling species' range shifts in a changing climate: The impacts of biotic interactions, dispersal distance and the rate of climate change. J Theor Bio 245: 59-65.

32. Hughes L (2003) Climate change and Australia: Trends, projections and impacts. Aus Eco 28: 423-443.

33. Kriticos DJ, Crossman ND, Ota N, Scott JK (2009) Climate change and invasive plants in South Australia. Canberra: CSIRO Climate Adaptation Flagship 97.

34. Walther GR, Post E, Convey P, Menzel A, Parmesan C and et al. (2002) Ecological responses to recent climate change. Nature 416: 389-395.

35. Kriticos D, Webber B, Leriche A, Ota N, Macadam I and et al. (2011) Global high-resolution historical and future scenario climate surfaces for bioclimatic modelling. Methods in Ecol Evo 3: 53-64.

36. Guisan A, Zimmerman NE (2000) Predictive habitat distribution models in ecology. Eco Model 135: 147-186.

37. Pearson RG, Dawson TP, Liu C (2004) Modelling species distributions in Britain: a hierarchical integration of climate and land-cover data. Eco 27: 285-298.

38. Soberón J (2007) Grinnellian and Eltonian niches and geographic distributions of species. Ecology Letters 10: 1115-1123.

39. Sutherst RW, Maywald G (1985) A computerized system for matching climates in ecology. Agri Eco Environ 13: 281-299.

40. Sutherst RW, Baker RH, Coakley SM, Harrington R, Kriticos DJ, et al. (2007) Pests under global change – meeting your future landlords? In: Candadell J, Pataki D, Pitelka L, editors. In Terrestrial ecosystems in a changing world. New York: Springer, Berlin Heidelberg. pp. 211-223.

41. Follak S, Strauss G (2010) Potential distribution and management of the invasive weed Solanum carolinense in Central Europe. Weed Res 50: 544-552.

42. Kriticos DJ (2006) Release Notes for Ozclim Australian Climate Change Scenarios for use in CLIMEX. CSIRO.

43. Poutsma J, Loomans AJM, Aukema B, Heijerman T (2008) Predicting the potential geographical distribution of the harlequin ladybird, Harmonia axyridis, using the CLIMEX model. BioControl 53: 103-125.

44. Senaratne K, Palmer WA, Sutherst RW (2006) Use of CLIMEX modelling to identify prospective areas for exploration to find

new biological control agents for prickly acacia. Aus J Entomol 45: 298-302.

45. Phillips SJ, Dudík M (2008) Modeling of species distributions with Maxent: new extensions and a comprehensive evaluation. Ecogra 31: 161-175.

46. Sutherst RW, Maywald G, Kriticos DJ (2007) CLIMEX Version 3: User's Guide. In: Ltd HSSP. Melbourne.

47. Webber BL, Yates CJ, Le Maitre DC, Scott JK, Kriticos DJ, et al. (2011) Modelling horses for novel climate courses: insights from projecting potential distributions of native and alien Australian acacias with correlative and mechanistic models. Divers and Distrib 17: 978-1000.

48. CLIMEX Version 3.0.2 (2007): Hearne Scientific Software

49. Rogers DJ, Reid RE, Rogers JJ, Addison SJ (2007) Prediction of the naturalisation potential and weediness risk of transgenic cotton in Australia. Agri Ecos & Environ 119: 177-189.

50. Saremi H, Kumar L, Sarmadian F, Heidari A, Shabani F (2011) GIS based evaluation of land suitability: a case study for major crops in Zanjan University region. J Food Agri Environ 9: 741-744.

51. Beaumont LJ, Hughes L, Poulsen M (2005) Predicting species distributions: use of climatic parameters in BIOCLIM and its impact on predictions of species' current and future distributions. Ecol Model 186: 251-270.

52. Guisan A, Zimmermann NE (2000) Predictive habitat distribution models in ecology. Ecol Model 135: 147-186.

53. Fitzpatrick MC, Weltzin JF, Sanders NJ, Dunn RR (2006) The biogeography of prediction error: why does the introduced range of the fire ant over-predict its native range? Global Ecol Biogeography 16: 24-33.

54. Earth Explorer (2012).

55. Global Biodiversity Information Facility (2012).

56. Missouri Botanical Garden (2012).

57. CAB Direct (2013).

58. Jain S, Al-Khayri J, Dennis V, Jameel M (2011) Date Palm Biotechnology. 1st ed: Springer 743.

59.   Woodcock I, Diana L (2010) Date Palm Plantation, Iran. 16: 26-29.

60.   Eshraghi P, Zarghami R, Mirabdulbaghi M (2005) Somatic embryogenesis in two Iranian date palm. African Journal of Biotechnology 4: 1309-1312.

61.   Shayesteh N, Marouf A (2010) Some biological characteristics of the Batrachedra amydraula Meyrick (Lepidoptera: Batrachedridae) on main varieties of dry and semi-dry date palm of Iran.

62.   Mahmoudi H, Hosseininia G (2008) Enhancing date palm processing, marketing and pest control through organic culture. J Org Sys 3: 30-39.

63.   Abbas I, Mouhi M, Al-Roubaie J, Hama N, El-Bahadli A (1991) Phomopsis phoenicola and Fusarium equiseti, new pathogens on date palm in Iraq. Mycological Res 95: 509.

64.   Auda H, Khalaf Z (1979) Studies on sprout inhibition of potatoes and onions and shelf-life extension of dates in Iraq. J Rad Phy Chem 14: 775-781.

65.   Heakal MS, Al-Awajy MH (1989) Long-term effects of irrigation and date-palm production on Torripsamments, Saudi Arabia. Geoderma 44: 261-273.

66.   Al-Senaidy M, Abdurrahman M, Mohammad A (2011) Purification and characterization of membrane-bound peroxidase from date palm leaves (Phoenix dactylifera L). Saudi J Bio Sci 18: 293-298.

67.   Markhand G (2000) Fruit characterization of Pakistani dates. Date Palm Research Institute.

68.   Hasan S, Baksh K, Ahmad Z, Maqbool A, Ahmed W (2006) Economics of Growing Date Palm in Punjab, Pakistan. Int J Agri and Bio 8: 1-5.

69.   Elhoumaizi M, Saaidi M, Oihabi A, Cilas C (2001) Phenotypic diversity of date-palm cultivars (Phoenix dactylifera L.) from Morocco. Genetic Res and Crop Evo 49: 483-490.

70.   Marqués J, Duran-Vila N, Daròs JA (2011) The Mn-binding proteins of the photosystem II oxygen-evolving complex are decreased in date palms affected by brittle leaf disease. Plant Phy and Biochem 49: 388-394.

71.   Sutherst R, Floyd RB (1999) Impacts of global change on pests, diseases and weeds in Australian temperate forests.

72.   Harrington R, Woiwod I, Sparks T (1999) Climate change and trophic interactions. Trends in Eco & amp Evol 14: 146-150.

73.   Kriticos DJ, Reynaud P, Baker RHA, Eyre D (2012) Estimating the global area of potential establishment for the western corn rootworm (Diabrotica virgiferavirgifera) under rain-fed and irrigated agriculture*. EPPO Bulletin 42: 56-64.

74.   Guisan A, Thuiller W (2005) Predicting species distribution: offering more than simple habitat models. Ecology Letters 8: 993-1009.

75.   Kriticos D, Potter K, Alexander N, Gibb A, Suckling D (2007) Using a pheromone lure survey to establish the native and potential distribution of an invasive Lepidopteran. J App Eco 44: 853-863.

76.   Gordon H, Rotstayn L, McGregor J, Dix M, Kowalczyk E, et al. (2002) The CSIRO Mk3 Climate System Model

77.   IPCC IPoCC (2007) Climate Change 2007: Synthesis Report. Summary for Policymakers.

78.   Houghton J, Ding Y, Griggs D, Noguer M, van der Linden P, et al. (2001) Climate change 2001: the science basis

79.   Jain S (2011) Prospects of in vitro conservation of date palm genetic diversity for sustainable production. Emirates J Food and Agri 23: 110-119.

80.   Burt J (2005) Growing date palms in Western Australia. 2-4.

81.   EEA Corine Land Cover 2000 seamless vector data

82.   NASA (2012) ASTER, Advance Spaceborn Thermal Emission and Reflection Radiometer California Institute of Technology.

83.   Kriticos DJ, Alexander NS, Kolomeitz S (2006) Predicting the potential geographic distribution of weeds in 2080. In: Preston C, Watts JH, Crossman ND, editors. 15th Australian Weeds Conference. Adelaide, Australia: Weed Management Society of South Australia 27-34.

84.   Crossman ND, Bryan BA, Cooke DA (2011) An invasive plant and climate change threat index for weed risk management: Integrating habitat distribution pattern and dispersal process. Eco Indic 11: 183-198.

85.   Svenning JC, Skov F (2007) Could the tree diversity pattern in Europe be generated by postglacial dispersal limitation? Ecol Lett 10: 453-460.

86. Chiew F, Kirono D, Kent D, Vaze J (2009) Assessment of rainfall simulations from global climate models and implications for climate change impact on runoff studies. 18th World IMACS Australia 3907-3914.

87. Suppiah R, Hennessy K (2007) Australian climate change projections derived from simulations performed for the IPCC 4th Assessment Report. 131-152 .

88. Suzaki T, Kume A, Ino Y (2005) Effects of slope and canopy trees on light conditions and biomass of dwarf bamboo under a coppice canopy. J Forest Res 10: 151-156

89. Basnet K (1992) Effect of topography on the pattern of trees in tabonuco (Dacryodes excelsa) dominated rain forest of Puerto Rico. Biotropica: 31-42.

90. Kravchenko AN, Bullock DG, Boast CW (2000) Joint multifractal analysis of crop yield and terrain slope. Agro J 92: 1279-1290.

91. Grattan S (2002) Irrigation water salinity and crop production: ANR Publications.

92. Tripler E, Shani U, Mualem Y, Ben-Gal A (2011) Long-term growth, water consumption and yield of date palm as a function of salinity. Agri Water Manage 99: 128-134.

Chapter 6

Organic Agricultural Practices among Small Holder Farmers in South Western Nigeria

Sijuwade Adebayo[1] and Idowu O Oladele[1]

[1]Department of Agricultural Economics and Extension, North West University Mafikeng Campus, South Africa

INTRODUCTION

Organic agriculture and biotechnology are two key innovations that are considered to have beneficial impacts on the future sustainability of agriculture (Wheeler, 2005). Conventional farming has played an important role in improving food and fibre productivity to meet human demands but has been largely dependent on intensive inputs of synthetic fertilizers and pesticides (Tu, Louws, Creamer, Mueller, Brownie, Fager, Bell and Shuijin, 2006). Moreover, the conventional intensive agricultural systems have side–effects which compromise food production in terms of quality and safety. Therefore, problems arising from conventional practices have led to the development and promotion of organic farming system that account of the environment and public health as main concerns (Melero, Ruiz Porras, Herencia and Madejon, 2005). Besides, traditional subsistence smallholding farming can no longer meet the needs and expectation of ever-increasing population of Nigeria (Adomi, Monday-Ogbomo and Inoni, 2003). Increasing agricultural productivity, self-sufficiency and poverty alleviation depend on the acceptance and full utilization of modern inputs, as long been recognized and policy formulation and implementation have been done (Aina 2007). The-Research-Extension-Farmers-Linkage-System (REFILS) has been able to ensure some awareness about the use of modern agro-inputs (Oladele, Sakagomi and Kazunobu 2006).

Organic farming represents a deliberate attempt to make the best use of local natural resources and is an environmental friendly system of farming. It relies much on ecosystem management which excludes external input, especially the synthetic ones. Ander son, Jolly and Green (2005) stated that organic farming is a production system that excludes the use of synthetically manufactured fertilizer, pesticides, growth regulators and livestock feed additives. The system relies on crop rotation, crop residues, animal manures, legumes, green manures, off-farm organic wastes, mechanical cultivation and aspects of biological pest control to maintain soil productivity and tilth, to supply plant nutrients and to control insects, weeds and other pests. According to Agbamu (2002), organic farming technology is frequently regarded as the solution to environmental problems that are related to agriculture as well as food safety. Furthermore, Conor (2004) pointed out that organic farming developed as a response to what was perceived to be polluting food supply by modern farming methods and the ensuing degradation of the environment with chemical and other by-products of the industry.

Soil quality is a necessary indicator of sustainability land. The two farming systems (organic and conventional) studied at farm level in Central Italy has emphasized interesting differences on soil quality. It became obvious that organic management affects soil microbiological and chemical properties by increasing soil nutrient availability, microbial biomass and microbial activity, which represent a set of sensitive indicators of soil quality. (Marinari, Mancinelli, Campiglia, Grego, 2006). The bacterial biomass that perform soil functions and resist environmental stress occurring under organic farms scores higher than in other farming systems (Mulder, De Zwart, Van Wijnen, Schouten, Breure, 2003). Furthermore, the results confirm the positive effects of organic manures and diversified crop rotations on soil quality aspects. Rigby and Ca'ceres (2001) and Defoer (2002) reported that organic agriculture tends to conserve soil fertility and system stability better than conventional farming systems. The Food and Agriculture Organization of the United Nations regards organic agriculture as an effective strategy for mitigating climate change and building robust soils that are better adapted to extreme weather conditions associated with climate changes (IFOAM, 2009; Pretty, 1999).

Organic agriculture promotes food safety and quality. The past decade has been characterized by escalating public concern towards nutrition and health and food safety issues (Crutchfield & Roberts, 2000). As a result, at present, consumers perceive relatively high risks associated with the consumption of conventionally grown produce compared with other public health hazards (Williams & Hammitt, 2000, 2001). Mitchell, Hong, Koli, Barrett, Bryant, Denison and Kaffka (2007) discovered that fruits and vegetables produced organically have increased levels of flavonoids which are reported to protect against cardiovascular disease (Hertog and Hollman, 1996) and to a lesser extent, against cancer (Knekt, Kumpulainen, Jarvinen, Rissanen, Heliovaara, Reunanen, Hakulinen and Aromaa, 2002) and other age-related diseases such as dementia (Commenges, Scotet, Renaud, Jacqmin-Gadda, Barberger-Gateau and Dartigues, 2000) whereas the levels of flavonoids did not vary significantly in conventional treatment. Furthermore, Lumpkin (2005), and Zug (2006) noted that the use of chemicals in vegetable production has been identified as a major source of health risk and a cause of extensive environmental damage.

Organic agriculture improves ecological health because farmers maintain nutrient balances in soil through locally available organic materials or recycled farm wastes (Park, Stabler and Jones, 2008; Hynes, 2009). Stolze, Piorr, Harring and Dabbert (2000) and Olsson et al (2001) concluded that nutrient balances on organic farms are often close to zero and that energy efficiency is found to be higher in organic farming than in conventional farming. It also encourages ecosystem service which sustains agricultural productivity and resilience and advocates production intensification through ecosystem management. Fertility management in organic farming relies on a long-term integrated approach rather than the more short-term much targeted solutions common in conventional agriculture (Watson et al., 2002). The practice of organic agriculture has been associated with returns on investment because it offers farmers a much more secure income than when they rely on only one or two inputs (Osborne, 2009; Mcguirk, 1990). Besides, organic farm precludes purchases of organic inputs, loans and thus the profit margin made by farmers increases and farmers are better off financially (Sanchez and Swaiminathan, 2005; Mei, Jewison and Greene, 2006).

Unlike organic agriculture, which emphasizes effective soil management and biodiversity, conventional agriculture (also referred to as intensive agriculture) relies on farming a single crop year after year. To overcome the imbalance imposed upon a conventional farm's ecosystem, harmful agents, such as pesticides and synthetic nitrogen fertilizers are used. Unfortunately, conventional agricultural practices exacerbate rather than alleviate the effects of climate change. The consequence of conventional farming's ecological imbalance is a decline in soil organic matter, soil structure, fertility, microbial and faunal biodiversity. Combine these impacts with the nutrient overload that ultimately ends up in waterways, deforestation, and overgrazing that occurs due to changes in land use, and it's not difficult to see why many are now stating that conventional agriculture represents an unsustainable long-term option.

The description of organic agriculture in the preceding section has led to the generation of research output recommended by Agricultural Knowledge and Information Systems (AKIS) in order to enhance organic agriculture and make it more sustainable and profitable. The information generated on organic agriculture by various AKIS has created the need for vegetable farmers to fill the information needs and bridge the gap in their production activities. The way in which information is sought is information seeking behaviour. The study attempts to analyse the information seeking behaviour and adoption of organic farming practices among vegetable farmers in South Western Nigeria.

ORGANIC AGRICULTURE

Organic agriculture is a holistic production management system which promotes and enhances agro-ecosystem health, including biodiversity, biological cycles, and soil biological activity. It emphasizes the use of management practices in preference to the use of off-farm inputs, taking into account that regional conditions require locally adapted systems. This is accomplished by using, where possible, agronomic, biological, and mechanical methods, as opposed to using synthetic materials, to fulfill any specific function within the system (FAO, 1999). The FAO/WHO Codex Alimentarius guidelines defined organic agriculture as "a holistic production management [whose] primary goal is to optimize the health and productivity of interdependent communities of soil life, plants, animals and people".

Similarly, the International Federation of Organic Agricultural Movements, with over 750 member organizations in 108 countries, defined it as "a whole system approach based upon sustainable ecosystems, safe food, good nutrition, animal welfare and social justice. Organic production therefore is more than a system of production that includes or excludes certain inputs (IFOAM, 2006; IFOAM, 2002). The aim of organic farming is to create integrated, humane, environmentally and economically viable agriculture systems in which maximum reliance is put on local or on-farm renewable resources, and the management of ecological and biological processes. The use of external inputs, whether inorganic or organic, is reduced as far as possible.

Certified organic food and fiber products are those that have been produced according to documented standards. They are foods that are guaranteed to have been produced and processed in a manner that avoids the use of synthetic fertilizers, pesticides, hormones, genetically modified organisms and irradiation, and which strives to enhance natural biological cycles and to meet minimum animal welfare standards.

"Certified organic agriculture" is defined as a certified system of agricultural production that seeks to promote and enhance ecosystem health while minimizing adverse effects on natural resources. It is seen not just as a modification of existing conventional practices, but as a restructuring of whole farm systems. However, "organic agriculture" is not limited to certified organic farms and products but can include all productive agricultural systems that use sustainable, natural processes, rather than external inputs, to enhance agricultural productivity (Scialabba and Hattam, 2002).

Organic farmers adopt practices to conserve resources, enhance biodiversity, and maintain the ecosystem for sustainable production and can lead to increased food production, in many cases we have seen a doubling of yields, which makes an important contribution to increasing the food security of a region (Park et al, 2008). Therefore, Non-certified organic agriculture' is defined as local, often traditional agriculture that is managed more or less in accordance with the principles of organic agriculture, but is not based on certification, trade and premium prices and it promises an alternative development path in rural areas of low-income countries (Halberg et al., 2006).

The principles of organic agriculture according to IFOAM are principle of Health-Organic agriculture should sustain and enhance the health of soil, plant, animal, human and planet as one indivisible; principle of ecology-organic agriculture should be based on living ecological systems and cycles, work with them, emulate them and help sustain them; principle of Fairness-Organic agriculture should build on relationships that ensure fairness with regard to the common environment and life opportunities and principle of care-organic agriculture should be managed in a precautionary and responsible manner to protect health and the well-being of current and future generations and the environment. Literature suggest that the farm, farmer and institutional factors drive farmers to adopt new technologies (De Francesco, Gatto, Runge and Tretini, 2008; Rehman, Mckemey,Yates, Cooke, Garforth, Tranter, Park and Dorward, 2007; Hattam, 2006). Factors such as the financial and social-economic impacts of new technologies, effects of new technologies on the risk of the farm, available resources and technology transfer programme also have an effect on the decision of the farmer to adopt new technologies.

Organic agriculture is fast emerging as the only sustainable long-term approach to food production. Its emphasis on recycling techniques, biodiversity, low external input and high level output strategies make it an ideal replacement for the petroleum intensive agricultural methods that are currently contributing to global warming (IFOAM, 2008; Swift et al, 2004). There are a number of factors indicating that organic agriculture is far more future proof than conventional agriculture. These include ecosystem services (Pimentel et al, 2005 and Stolze et al, 2000); Ecological health (Backer et al, 2009, D'Agostino and Sovacool, 2011); Soil fertility and system stability (Reddy, 2010, Mader et al, 2002); mitigating climate change (FiBL, 2007, Lee, 2005); food safety and quality (Gallagher et al, 2005, Makatouni, 2002; Magnusson et al, 2001 and Torjusen et al, 2001); return on investment and poverty alleviation (Rigby and Caceres, 2001); consumer preferences (Willer and Youssefi, 2007, Chen, 2007 and Mondelaers et al, 2009); value addition (Ohmart, 2003, Mitchell et al, 2007); market niche (Alroe and Noe, 2008) and indigenous knowledge (Tengo and Belfrage, 2004, IFOAM, 2003).

METHODOLOGY

The area of study is southwestern Nigeria which comprises of six states namely: Oyo, Osun, Ogun, Ondo, Ekiti and Lagos States. Southwest is situated mainly in the Tropical Rainforest Zone, though with swamp forest in the coastal regions in Lagos, Ogun, Ondo and Delta States. The agricultural sector forms the base of the overall development thrust of the zone. The zone covers an area ranging from swamp forest to western up lands, in between are rain forest and the northern parts of Oyo and Ogun states having derived Guinea savannah vegetation. The areas lie between latitude 5 degrees and 9 degrees North and longitude 2 degrees and 8 degrees East. It is bounded by the Atlantic Ocean in the south, Kwara and Kogi states in the north, Eastern Nigeria in the east and Republic of Benin in the west. It has a land area of about 114,271km square representing 12% of the country's total land areas. The high concentration of agricultural activities justifies the choice of the study area (NARP, 1996).

The research design of the study is descriptive and quantitative which is defined by Bless and Higson-Smith (2000), as a study concerned with the condition that exist, practices that prevail, beliefs and attitudes that are held, processes that are on-going and trends that are developing. The study profile organic farming practices in southwestern Nigeria. The population of the study is the entire population of vegetable farmers in the South Western Nigeria. Cluster sampling technique was adopted for selecting the required sample of urban vegetable producers. From literature and preliminary surveys, vegetable production in urban areas that is market oriented is mostly carried out along perennial sources of water or lowlands. This constrains farmers to clusters around these sources of water. Therefore, cluster sampling is considered appropriate. The sampling technique involves random selection of three states in the southwestern Nigeria which were Oyo, Ogun and Ondo. Three local government areas in the urban were selected from each state to give a total number of nine local government areas used for the study. The choice of these Local government areas is based on the dominance of vegetable producers in the different areas. The three local government areas chosen in Oyo state were Akinyele, Egbeda and Ogbomoso south. The three local government areas chosen in Ogun state were Odeda, Obafemi Owode and Abeokuta north. The three local government areas selected in Ondo state were Akure south, Akure north and Ifedore. A cluster of vegetable producers was selected from each of the local government areas to give total of nine clusters. Fifty producers were randomly selected from each of the nine clusters to give a total sample size of four hundred and fifty respondents for the study.

Data for this study was generated from primary sources based on the objective of the study. Interview schedule was used to elicit information from the respondents. The questionnaire consisted of 14 organic farming practices in southwestern Nigeria from which the respondents indicated use and non-use. These practices are crop rotation, application of compost, mulching of crops, inter cropping, mixed cropping, crop residues, cover crop, animal manure, organic fertilizer, bio control, natural insect predator. A split half technique was used to determine the reliability coefficient with a reliability coefficient of 0.85. The questionnaire was face validated by panel of experts on agricultural extension, agronomist and organic agricultural researcher. The panel consisted of lecturers in agricultural extension and Agronomy. The study took into account the ethical consideration which was addressed through, voluntary participation. Data were analyzed with the Statistical Package for Social Sciences (SPSS) 18.0 using means and standard deviation.

RESULTS AND DISCUSSIONS

Table 1 shows a list of 14 organic agriculture practices from which the respondents were asked to indicate their use or otherwise using a 2 ponit scale of Yes (2) and No(1). The actual mean is 1.5 due to the rating scale and a mean of greater than 1.5 denoted a use while a mean less than 1.5 denoted non-use. The mean scores of 11 out of 14 practices were above the actual mean which implies the use of these organic agriculture practices. These technologies are: minimum tillage, crop rotation, sanitation, intercropping, green manure, cover crop, fire, compositing, organic fertilizer, animal manure, and mulching. The results revealed the most prominent organic agriculture practices were minimum tillage (1.81, SD=0.9); crop rotation (1.80, SD=0.7) and mulching (1.79, SD=0.6). With respect to the use of minimum tillage, it is the practice that minimises the disturbance of the soil. The soil is not tilled intensively thereby improving the soil structure. It is a cultivation operation whereby soil is disturbed as little as possible to produce crop. Mulch residue from the previous crop is left on the soil surface which aids in retarding weed growth, conserving moisture, and controlling erosion. Therefore, the practice of minimum tillage is a common operation among the farmers that is usually carried out in order to prepare the soil before planting exercise. Baldwin (2006) noted that many organic farmers typically manage weeds mechanically and, therefore, cannot focus on building soil structure in the same way as conservational tillage practitioners which often relies on herbicides for weed control. Instead, organic farmers use innovative practices such as crop rotations, green manuring, and biological pest control to improve the soil structure and conserve soil organic carbon.

Crop rotation as one of the practices can be attributed to the use of indigenous knowledge, where farmers' belief that soil needs rest and some measure should be put in place to ensure soil maintenance and fertility. One of such measures is bush fallowing whereby a farmland that have been cultivated for some number of years is left uncultivated for few years in order to fallow and regain its lost nutrients. Crop rotation is another measure that is used by the farmers for this purpose. In this case, the farm land is not abandoned but crops that are cultivated on the farm are planted in sequence in order to maintain the soil fertility. Crop rotation is a practice that is as old as farming practice itself. Subba Rao (1999) and Stockdale, et al (2000) observed that crop rotations and varieties are selected to suit local conditions having the potential to sufficiently balance the nitrogen demand of crops. Furthermore,Bending and Lincoln, (1999) in their work among the US farmers noted that organic growers commonly plant rapeseed, mustard, and other brassicas as rotation crops to 'clean up' soil during the winter months. Besides, Crop rotations comprising both grass-clover fields and arable crops have shown to be relatively robust in relation to most problems with weeds, pests and diseases (Dubois et al, 1999).

Mulching ranks highly as a cost-effective means of crop residue usage against soil erosion in annual row-cropping systems on sloping lands; and is at the centre of a resurgent soil conservation ethic in much of North America (Shelton et al., 1995). However, it is not commonly used among the vegetable farmers who reported that mulching is predominantly used by yam producers. The findings of Junge et al, (2009) showed that mulching and cover cropping were mostly regarded as not labour-intensive, highly cost-effective, compatible and easy and cheap to adopt. The farmers had a positive impression of the effectiveness as erosion control measures and also mentioned additional advantages, such as the increased soil fertility from the decomposition of organic material and the release of nutrients however disadvantage of mulching was seen in the amount of grass required, the main material used as mulch in the area.

Table 1: Distribution of the respondents by use of organic agricultural practices

Organic agriculture practices	Mean	SD
Minimum tillage	1.81	0.9
Crop rotation	1.80	0.7
Farm Sanitation	1.69	0.8
Intercropping	1.66	0.2
Green manure	1.60	0.9
Cover crop	1.55	0.8
Fire	1.53	0.6
Composting	1.60	0.4

Organic fertilizer	1.68	0.9
Animal manure	1.71	0.3
Mulching	1.79	0.6
Natural pesticides	0.36	0.6
Farm scaping	0.16	0.6
Bio control	0.13	0.3

Other organic agricultural practices used by farmers include practices. Farm sanitation (1.69, SD=0.8), intercropping (1.66, SD=0.2), green manure (1.60, SD=0.9) and cover crop (1.55, SD=0.8). Farm Sanitation is keeping the field clean which help in preventing the growth and multiplication of weed, pest and diseases. The reason may be because farmers are also aware of those things that can prevent them from having good yield or output. Farmers go to farm everyday even after the planting period to weed at interval, remove any form of crop residue or decay of dead animal on their farm that can attract pests and diseases to the crop planted and can cause pollution in the environment. Farmers are aware that if weed are left to grow on their plot, it will compete with the crop planted for the available nutrients and will reduce their yield during harvest. Besides, some weeds affect the crop leaving a residual effect on the crop which can affect the taste or the appearance of the crop. Whenever this happened, the farmer will run at a loss because such crop will not attract buyers and may have to be sold at a ridiculous price.

Baumann et al., (2000) showed that intercropping as a cultural method can be used to suppress weeds and reduces pest population because of the diversity of crops grown. According to Sullivan (2003), if susceptible plants are separated by non-host plants that can act as a physical barrier to the pest, the susceptible plant will suffer less damage. Furthermore, intercropping reduced the nitrate content in the soil profile as intercropping uses soil nutrients more efficiently than sole cropping (Zhang and Long Li, 2003).

Katyal (2000) reported the application of organic manure as the only option to improve the soil organic carbon for sustenance of soil quality and future agricultural productivity. Wambani et al. (2006) compared the effect of farmyard manure application with recommended rate of inorganic fertilizer and it was discovered that the recommended rate of organic manure was the most profitable and preferred by the farmers because of their low cost, availability of organic manure and longer persistence of kales under these treatments.

Cremer et al; (1996) showed that cover crop residues interfere with the emergence of weed through the allelopathic effect. In addition, Langdale et al. (1991) concluded that cover crops reduced soil erosion by 62 per cent based on a comparison of bare soil and soil planted with a cover crop in the south eastern United States. Results presented for the use of Tithonia and legume cover crops shows increase grain yields significantly in Eastern Uganda (Delve and Jama, 2002). Moreover, Cover crops can improve soil quality (Dabney et al. 2001), and when planted at the beginning of the transition phase, may provide essential soil-building properties and improve weed suppression (Barberi 2002; Martini et al. 2004); however, soil quality effects and ability of cover crops to suppress weed species varies among cover crop species (Melander et al, 2005; Snapp et al, 2005).

The results further shows that the use of organic agricultural practice covered fire (1.53, SD=0.6), composting (1.60, SD=0.4), organic fertilizer (1.68, SD=0.9) and animal manure (1.71, SD=0.3). Wilson (2007) found that flame weeding also called flame cultivation or flaming, is a thermal physical control method that is part of the National Organic Program (NOP) under the organic foods production act of 1990. Flame weeding delays the presence of weeds in crop beds by killing the weeds present before the crop has breached the soil. This can significantly reduce hand-weeding labor costs. Farmers see the use of fire as an easy and faster method of clearing the weeds, trees and bushes particularly at the on-set of planting season when the land is prepared. Besides, some farmers believe that when the land is prepared with fire, the ash of the weeds, trees or residues that were burnt will make the soil to be fertile. Farmers see the use of fire for clearing as cost-effective compared to the use of hired labour. Anon (1999) reported that in Iowa, farmer feedback on flame weeding has been positive however burning as labour-saving tool to clear land and to prevent weed infestation is now being brought into question and many development agencies now advocate no-burning. In the communities, however, it is less a question of burning or no-burning but rather when, where, and how to reduce its negative impact (Aalangdong et al., 1999). Some northern farmers have made a conscious decision to cease bush burning with the aim of regenerating organic material (Millar et al 1996). Singh (2003) noted that organic farmers in India reported the capacity of manure (compost) to fulfil nutrient demand of crops adequately and promote the activity of beneficial macro-and micro-flora in the soil. Also, Ouédraogo et al (2001) showed that farmer was aware of the role of compost in sustaining yield and improving soil quality. However, lack of equipment and adequate organic material for making compost, land tenure and the intensive labour required for making compost are major constraints for the adoption of compost technology. Olayide et al (2011) assessing farm-level limitations and potentials for organic agriculture in northern Nigeria, discovered that the current levels of organic fertilizer use as share of the minimum requirements for take-off for organic agriculture in Nigeria was low despite its potentials.

Vanlauwe, (2004) noted that livestock manure is important in maintaining soil organic matter levels, a critical factor in soil health. Additionally, Omiti et al, (1999) noted that animal manure compost is the most common source of soil amendment in organic agriculture in Nigeria and indeed Africa. Farmers are fully aware of the fertilizing value of animal manure as well as the differences, for example, in nutrient release between the manures as also reported by Dittoh (1999) and Karbo et al. (1999). However, Mafongoya et al (2006) reported that in Africa, though, animal manure is one of the mostly used organic inputs, but as the need for increased agricultural production rises; it has been found to be limited in quality and quantity. Williams (1999) reported similar result among farmers in semi-arid West Africa.

However the use of natural pesticides (0.36, SD=0.6), farm scaping (0.16, SD=0.6) and Bio control (0.13, SD=0.3) were below the actual mean which indicate non-use by the farmers. This may be because these practices do not fit in to the farming system in the study areas. It can also be attributed to the technicality of the use of these practices usch that the application of the practices and the associated legislation and the process of securing permission for the use of these practices.

CONCLUSIONS

The paper has shown the nature and trend of the use organic agricultural practices among smallholder farmers in South Western Nigeria by highlighting organic agricultural practices that are prominent and those that were less prominent as well as practices that are not in use. Due to the prevailing opportunities and benefits associated with the use of these practices, this paper recommend that farmers should increase their awareness and use of organic agricultural practices.

REFERENCES

1. Aalangdong, O, I., Kombiok, J, M., & Salifu, A, Z., 1999, Assessment of non-burning and organic-manuring practices, ILEIA newsletter, 15(1/2): 47–48

2. Adomi, E, E; Monday,- Ogbomo, O., & O.E .Inoni, O, E., 2003, Gender factors in crop farmers' access to agricultural information in rural areas of Delta State, Nigeria, *library Review* 52(8):388-93.

3. Agbamu, J, U., 2002, Agricultural Research Extension Farmer Linkages in Japan: "Policy issues for Sustainable Agricultural Development in Developing Countries" *International Journal of Social and Policy Issues,* 2002(1): 252-263.

4. Aina, L, O., 2007, Globalization and small-scale farming in Africa: What role for information centers. World Library and Information Congress: 73rd IFLA General Conference and Council, Durban, South Africa, August 19-23, 2007. Accessed February 6, 2010. http://www.ifla.org/iv/ifla73/index.html.

5. Alrøe, H, F., & Noe, E., 2008, What Makes Organic Agriculture Move: Protest, Meaning or Market? A Polyocular Approach to the Dynamics and Governance of Organic Agriculture, *International Journal of Agricultural Resources, Governance and Ecology,* 7, (1/2), 2008

6. Anderson, J, B., Jolly, D, A., & Green, R., 2005, Determinants of farmer adoption of organic production methods in the fresh-market produce sector in California: A logistic regression analysis. A paper presented at the Western Agricultural Economics Association 2005 Annual Meeting, July 6-8, 2005, San Francisco, California. http:// ageconsearch.umn.edu/bitstream/36319/1/sp05an01.pdf, accessed July 2011.

7. Backer, E, D., Aertsens, J., Vergucht, S., & Steurbaut, W., 2009, Assessing the ecological soundness of organic and conventional agriculture by means of life cycle assessment. *British Food Journal* 111 (10):1028-1061.

8.   Baldwin, K, R., 2006, Organic Production- Conservation of Tillage on Organic Farms, Published by North Carolina Cooperative Extension Service.

9.   Barberi, P., 2002, Weed management in organic agriculture: are we addressing the right issues? *Weed Response* 42:177–193.

10.   Baumann, D, T., Bastiaans, L., & Kropff, M, J., 2000, Competition and Crop Performance in a Leek–Celery Intercropping System, *Crop Science* 41:764–774 (2001).

11.   Bending, G, D., & Lincoln, S, D., 1999, Characterization of volatile sulphur containing compounds produced during decomposition of Brassica juncea tissues in soil, *Soil Biology and Biochemistry*, 31: 695-703

12.   Bless, C., & Higson-Smith, C., 2000, Fundamentals of social research methods: An African Perspective, 3rd Edition, Juta Education (Pty) Ltd, Cape Town, pp.37-42

13.   Chen, M, F., 2007, "Consumer attitudes and purchase intentions in relation to organic foods in Taiwan: moderating effects of food-related personality traits", *Food Quality and Preference,* 18 (7): 1008-21.

14.   Commenges, D., Scotet, V., Renaud, S., Jacqmin-Gadda, H., Barberger-Gateau, P., & Dartigues, J, F., 2000, Intake of flavonoids and risk of dementia, *European Journal of Epidemiology* 2000, *16*, 357-363

15.   Connor, J, O., 2004, Organic Matter ,Bi-monthly Magazine of Irish Organic farmers and a growers association West Cork, waterfall, Beara. 2(14)

16.   Creamer, N, G., Bennett, M, A., Stinner, B, R., Cardina J., & Regnier, E, E., 1996, Mechanisms of weed suppression in cover crop-based production systems, *Horticultural Science,* 31:410-413

17.   Crutchfield, S., Buzby J., Frenzen P., Allshouse J., & Roberts D., 2000, The Economics of Food Safety and International Trade in Food Products, United States Department of Agriculture Economic Research Service 1800 M Street NW Washington, DC, 20036

18.   D'agostino, A, L., & Sovacool, B, K., 2011, Sewing climate-resilient seeds: implementing climate change adaptation best practices in rural Cambodia, Mitigation Adaptive Strategy Global Change

DOI 10.1007/s11027-011-9289-7 Springer Science+Business Media B.V. 2011

19. Dabney, S, M., Delgado, J, A., & Reeves, D, W., 2001, Using winter cover crops to improve soil and water quality, *Communication Soil Science Plant Analysis* 32:1221–1250.

20. De Francesco, E., Gatto, P., Runge, F., & Trestini, S., 2008, Factors affecting farmers' participation in agri-environmental measures: A Northern Italian perspective, *Journal of Agricultural Economics*, 59: 114- 131.

21. Defoer, T., 2002, Learning about methodology development for integrated soil fertility management, *Agricultural Systems*, 73: 57–81

22. Delve, R, J., & Jama, B., 2002, Developing organic resource management options with farmers in eastern Uganda, Proceedings of the 17th World Congress of Soil Science, Bangkok, Thailand, 2002

23. Dittoh, S., 1999, Sustainable soil fertility management: Lessons from action research.Ileia newsletter 15(1/2): 51–52

24. Dubois, D., Gunst, L., Fried, P., Stauffer, W., Spiess, E., Mader, P., Alfoldi, T., Fliebbach, A., Frei. R., & Niggli, U., 1999, Dok-Versuch: Ertragsentwicklung und Energieeffizienz. Agrarforschung 6, 71- 74

25. FAO, 1999, Organic Agriculture, Food and Agriculture Organization of the United Nations, Rome, <http://www.fao. org/unfao/bodies/COAG/COAG15/X0075E.htm>. Accessed [26 February 1999]

26. Gallagher, K., Ooi, P., Mew, T., Borromeo, E., Kenmore, P., & Ketelaar, J, W., 2005, Ecological basis for low-toxicity integrated pest management (IPM) in rice and vegetables, In The pesticide detox (ed. J. Pretty), London,UK: Earthscan pp.116-134, 294pp.

27. Hazlberg, N., Alrøe, H, F., Knudsen, M, T., & Kristensen, E, S., 2006, Synthesis: prospects for organic agriculture in a global context, CAB International 2006. *Global* Development of Organic Agriculture: Challenges and Prospects pp.343-357

28. Hattam, C., 2006, Adopting certified organic production: evidence from small-scale avocado producers in Michoacaan, Mexico, Unpublished PhD Thesis, University of Reading.

29. Hertog, M, G, L., & Hollman, P, C, H., 1996, Potential health effects of the dietary flavonol quercetin, *European Journal of Clinical Nutrition* 1996,*50*, 63- 71

30. IFOAM, 2002, IFOAM – Norms for Organic Production and Processing, International Federation of Organic Agriculture Movements, Bonn (www.ifoam.org)

31. IFOAM, 2009, Global Organic Agriculture: Continued Growth. BioFacch World Organic Trade Fair 2009 in Nurenberg, Germany

32. IFOAM, 2003, Organic and Like-Minded Movement in Africa, International Federation of Organic Agriculture Movements (IFOAM), Bonn, 2003: 102-108

33. Junge, B., Deji, O., Abaidoo, R., Chikoye, D., & Stahr, K., 2009, Farmers' Adoption of Soil Conservation Technologies: A Case Study from Osun State, Nigeria, *The Journal of Agricultural Education and Extension,* 15:3, 257-274

34. Karbo, N., Bruce, J., & Otchere, E, O., 1999, The role of livestock in sustaining soil fertility in northern Ghana, ILEIA newsletter 15(1/2): 49–50

35. Katyal, J. C., 2000, Organic matter maintenance: Mainstay of soil quality, *Journal of the Indian Society of Soil Science*, 2000, 48, 704–716.

36. Langdale, G. W., Blevins R. L., Karlen D. L Mccool K.K., Nearing M.A, Skidmore E.L., Thomas A.W., Tyler D.D., & Williams J.R. 1991, Cover crop effects on soil erosion by wind and water, In W.L. Hargrove (Ed.), *Cover Crops for Clean Water,* Pp. 15-22, Soil and Water Conservation Society, Ankeny, IA.

37. Lee, D, R., 2005, Agricultural sustainability and technology adoption: Issues and policies for developing countries, *American Journal of Agricultural Economics* 87 (5):1325-1334

38. Lumpkin, H., 2005, Organic Vegetable Production: A Theme for International Agricultural Research. Seminar on production and export of organic fruit and vegetables in Asia, FAO corporate Document Repository, http:www.fao.org/DOCREP/006/AD429E/ad429e13.htm

39. Mader, P., Fliessbach, A., Dubois, D., Gunst, L., Fried, P., & Niggli, U., 2002, "The ins and outs of organic farming", *Science*, 298 (5600): 1889-90.

40. Mafongoya, P, L., Bationo, A., Kihara, J., Waswa, B, S., 2006, Appropriate technologies to replenish Soil fertility in Southern Africa, *Nutrient Cycling in Agro ecosystem* 76: 127-151

41. Magnusson, M, K., Arvola, A., & Koivisto-Hursti, U, K., 2001, Attitudes towards organic foods among Swedish consumers, *British Food Journal*, 103:209– 226

42. Makatouni, A., 2002, What motivates consumers to buy organic food in the UK?: Results from a qualitative study, *British Food Journal,* 104:345–352.

43. Marinari, S., Mancinelli, R., Campiglia, E., & Grego, S., 2006, Chemical and biological indicators of soil quality in organic and conventional farming systems in Central Italy, *Ecological Indicators* 6 (2006) 701–711

44. Martini, E, A., Buyer, J, S., Bryant, D, C., Hartz, T, K., & Denison, R, F., 2004, Yield increases during the organic transition: improving soil quality or increasing experience? *Field Crops Research* 86:255–266.

45. Mei, Y., Jewison, M., & Reene, C., 2006, Organic products market in China, USADA Foreign Agricultural Service, GAIN Report, CH6405, June

46. Melander, B., Rasmussen, I, A., & Barberi, P., 2005, Integrating physical and cultural methods of weed control—examples from European research,*Weed Science* 53:369–381

47. Melero, S., Ruiz Porras, J, C., Herencia, J, F., & Madejon, E., 2005, Chemical and biochemical properties in a silty loam soil under conventional and organic management, *Soil & Tillage Research* 90 (2006) 162–170

48. Millar, D., Ayariga, R., & Anamoh, B., 1996, Grandfather's way of doing: gender relations and the yaba-itgo system in Upper East Region, Ghana. In: Reij C, Scoones I and Toulmin C (eds) Sustaining the soil, Indigenous soil and water conservation in Africa, pp 117-125. London: Earthscan Publications

49. Mitchell, A, E., Hong, Y, J., Koh, E., Barrett, D, M., Bryant, D, E., Denison, R, F., & Kaffka, S., 2007, Ten-Year Comparison of the Influence of Organic and Conventional Crop Management Practices on the Content of Flavonoids in Tomatoes, *Journal of Agricultural and Food Chemistry*2007, 55, 6154-6159

50. Mulder, C, H., De Zwart, D., Van Wijnen, H, J., Schouten, A, J., & Breure, A, M., 2003, Observational and simulated evidence of ecological shifts within the soil nematode community of agro ecosystems under conventional and organic farming, *Functional Ecology* 17 (4): 516-525.

51. NARP, 1996, Staff Appraisal Report, National Agricultural Research Project: Newman, and Newman, J. (1985). Information work: the new divouris.*British Journals of Sociology*, 36 (4): 497-515.

52. Ohmart, J, L., 2003, "Direct Marketing with Value-added products (or: "Give me the biggest one of those berry tarts!")", University of California Sustainable Agriculture Research and Education Program.

53. Oladele, O, I., Jun-Ichi, S., & Kazunobu, T., 2006, Research – extension-farmer-linkage system in South western Nigeria, *Journal of Food, Agriculture & Environment* 4(1):99-102

54. Olayide, O, E., Anthony, E, I., Arega, D, A., & Vincent, A., 2011, Assessing Farm-level limitations and Potentials for Organic Agriculture by Agro-ecological Zones and Development Domains in Northern Nigeria of West Africa, *Journal of Human Ecology*, 34(2): 75-85 (2011)

55. Olsson, P., & Folke, P., 2001, Local ecological knowledge and Institutional dynamics for ecosystem management, Study of Lake Racken water shed, Sweden, *Ecosystems* 4: 85-104

56. Omiti, J, M., Freeman, H, A., Kaguongo, W., Bett, C., 1999, Soil Fertility Maintenance in Eastern Kenya: Current Practices, Constraints, and Opportunities, CARMASAK Working Paper No. 1. KARI/ICRISAT, Kenya

57. Osborne, B., 2009, Organic farming, Encarta encyclopaedia

58. Ouédraogo, E., Mando, A., & Zombré, N, P., 2001, Use of compost to improve soil properties and crop productivity under low input agricultural system in West Africa, *Agriculture, Ecosystems & Environment*, 84 (3): 259-266.

59. Park, Stabler, Jones, 2008, Evaluating the role of environmental quality in the sustainable rural economic development of England, *Environment, Development And Sustainability* .10, 69-88

60. Pretty, J., 1999, Can sustainable agriculture feed Africa? New evidence on progress, processes and impacts, *Environment, Development and Sustainability* 1: 253–274.

61. Reddy, B, S., 2010, Organic Farming: Status, Issues and Prospects-A Review. *Agricultural Economics Research Review* Vol. 23 July-December 2010 pp 343-358

62. Rehman, T., McKemey, K., Yates, C.M., Cooke, R.J., Garforth, C.J., Tranter, R.B., Park, J.R., and Dorward, P.T 2007. Identifying and understanding factors influencing the uptake of new technologies on dairy farms in SW England using the theory of reasoned action, *Agricultural Systems*, 94: 287- 290.

63. Rigby, D., & Caceres, D., 2001, "Organic farming and the sustainability of agricultural systems", *Agricultural Systems*, 68 (1): 21-40.

64. Sanche, P, A., & Swaminathan, M, S., 2005, Hunger in Africa: The link between unhealthy people and unhealthy soils, The lancets 365:442-444

65. Scialabba, N., 2000, Factors Influencing Organic Agriculture Policies with a focus on Developing Countries, IFOAM 2000 Scientific Conference, Basel, Switzerland, 28-31 August 2000. 13p.

66. Shelton, D, P., Dickey, E, C., Hachman, S, D., Steven, D., & Fairbanks, K, D., 1995, Corn residue cover on soil surface after planting for various tillage and planting systems, *Journal of Soil Water Conservation* 50, 399–404.

67. Singh, S., & George, R., 2012, Organic Farming: Awareness and Beliefs of Farmers in Uttarakhand, India, *Journal of Human Ecology*, 37(2): 139-149 (2012)

68. Snapp, S, S., Swinton, S, M., Labarta, R., Mutch, D., Black, J, R., Leep, R., Nyiraneza, J., & O'Neil, K., 2005, Evaluating cover crops for benefits, costs and performance within cropping system niches, *Agronomy Journal*, 97:322–332.

69. Stockdale, E., et al., 2000, Agronomic and environmental implications of organic farming systems, *Advanced Agronomy*, 2000, 70, 261–327

70. Stolze, M., Piorr, A., Ha°Ring, A., & Dabbert, S. 2000, "The environmental impact of organic farming in Europe", Organic

Farming in Europe,*Economics and Policy*, 6: 23-86 University of Hohenheim, Hohenheim

71. Subba Rao, I, V., 1999, Soil and environmental pollution – A threat to sustainable agriculture, *Journal of Indian Society of Soil Science*, 1999, 47, 611–633.

72. Sullivan, P., 2003, Intercropping principles and production practices. Agronomy systems guide, ATTRA (Appropriate Technology Transfer to Rural Areas), 12 pp (http:// www.attra.ncat.org)

73. Tengo, M., & Belfrage, K., 2004, Local management practices for dealing with change and uncertainty: a cross-scale comparison of cases in Sweden and Tanzania. *Ecology and Society*, 9(3):4, 22p. Available at www.ecologyandsociety.org/vol9/iss3/art4

74. Torjusen, H., Lieblein, G., Wandel, M., & Francis, C, A., 2001, Food system orientation and quality perception among consumers and producers of organic food in Hedmark County, Norway, *Food Quality and Preference* 12:207–216.

75. Tu, C., Louws, F, J., Creamer, N, G., Mueller, J, P., Brownie, C., Fager, K., Bell, M., & Shuijin, Hu, 2006, Responses of soil microbial biomass and N availability to transition strategies from conventional to organic farming systems, *Agriculture, Ecosystems and Environment* 113 (2006) 206–215

76. Vanlauwe, B., 2004, Integrated soil fertility management research at TSBF: the framework, the principles, and their application. In: Bationo, A. (Ed.), Managing Nutrient Cycles to Sustain Soil Fertility in Sub-Saharan Africa, Academy Science Publishers, Nairobi.

77. Watson, C, A., Younie, D., Stockdale, E.A., Cormack, W, F., 2000, Yields and nutrient balances in stocked and stockless organic rotations in the UK, Aspects *Applied Biology* 62, 261–268.

78. Wheeler, S., 2005, Factors Influencing Agricultural Professionals' Attitudes Towards Organic Agriculture and Biotechnology

79. Willer, H. and Youssefi, M. 2007, The World of Organic Agriculture – Statistics and Emerging Trends, International Federation of Organic Agriculture Movements (IFOAM), Germany and Research Institute of Organic Agriculture FiBL, Bonn. 77p.

80. Williams, T, O., 1999, Factors influencing manure application by farmers in semi-arid west Africa, *Nutrient Cycling in Agro ecosystems* 55: 15–22, 1999.

81. Williams, P, R., & Hammitt, J, K., 2001, Perceived risks of conventional and organic Produce: Pesticides, Pathogens and Natural toxins, *Risk Analysis* 21 (2): 319-330.

82. Williams, P, R, D., Hammitt, J, K., 2000, A comparison of organic and conventional fresh produce buyers in Boston Area, *Risk Analysis* 20 (5), 735–746.

83. Zhang, F., & Long Li, 2003, Using competitive and facilitative interactions in intercropping systems enhances crop productivity and nutrient-use efficiency, *Plant and Soil* 248: 305–312, 2003.

84. Zug, S., 2006, Monga—seasonal food insecurity in Bangladesh—Bringing the information together, The Journal of Social Studies, 111(July–Sept. 2006)

Chapter 7

Beneficial and Negative Impacts on Soil by the Reuse of Treated/Untreated Municipal Wastewater for Agricultural Irrigation – A Review of the Current Knowledge and Future Perspectives

Juan C. Durán–Álvarez[1] and Blanca Jiménez–Cisneros[2]

[1]Centre of Applied Science and Technological Development, National Autonomous University of Mexico, Mexico D.F., Mexico
[2]International Hydrological Programme (IHP), UNESCO, Paris, France

INTRODUCTION

The scarcity of water for human use, such as food and energy production, manufacturing, drinking water and ecosystem conservation is a global problem for which the solution goes beyond merely the preservation of freshwater sources [1–2]. Although three quarters of the Earth´s surface is covered by water, most of this water is either contained in oceans or confined in glaciers [3]. The volume of freshwater available for human activities (less than 1%) is unequally distributed throughout the globe; in some cases this water is confined to the deep sub–soil or is polluted [4]. Furthermore, the desertification of large areas caused by climate change has intensified the lack of water sources in cities and rural areas throughout the world [5]. Water scarcity results in food scarcity, since 70% of the water withdrawn for human activities goes to agriculture [6]. In zones where rain–fed agriculture is practiced, decay in crop yields is observed when droughts occur, which results not only in the scarcity of food but also the decrease in incomes due to falling crop sales [7]. The use of freshwater for agricultural irrigation limits the volume of freshwater available for human consumption; therefore, recycling of water becomes necessary for agricultural irrigation in dry zones. The idea of reusing wastewater to irrigate is not new; it actually originated around 3000 B.C. People in these ancient civilizations knew that wastewater contained both water and compounds that benefited the soil and thus they used it in a planned way to increase crop yields [8].

Commonly, reusing wastewater in agriculture is considered a deleterious practice since it may introduce pollutants to the environment, spread waterborne diseases, generate odor problems and result in aversion to the crops. Nevertheless, this kind of reuse may result in some benefits for soils, crops and farmers. Nowadays, the reuse of wastewater in agriculture is seen in some countries as a convenient environmental strategy [9–10]; municipal wastewater is therefore considered an appropriate option for reuse. This kind of wastewater contains a significant load of biodegradable organic material (carbon and nitrogen) as well as most of the mineral macronutrients (e.g. phosphorous, potassium, magnesium and boron) and micronutrients (e.g. molybdenum, selenium and copper) which are necessary for the growth of crops. Accumulation of organic matter in soil by irrigation with wastewater can be beneficial as it may result in the enhancement of the physical structure of the soil, the increase in the soil microbial activity and the improvement of soil performance as a filter and degrading media for pollutants. Conversely, a fraction of the organic matter contained in wastewater is due to the occurrence of organic pollutants (e.g. polyaromatic hydrocarbons and polychlorinated biphenyls) and pathogenic microbial agents [11–12]. Because of the presence of organic, inorganic and microbial pollutants in wastewater, a prior step of depuration is necessary before reuse in irrigation in order to avoid the pollution of soil, crops and the nearby water sources, and thus the dissemination of waterborne diseases or the degradation of soil. The extent at which wastewater has to be treated prior to irrigation depends on the restrictions established in local or international water quality criteria for irrigation [13]. Primary treatment schemes (coagulation–flocculation with sedimentation or aerobic/ anaerobic stabilization pounds) are used for treating wastewater to irrigate crops that are not intended for human consumption (e.g. fodder), while secondary treatment of wastewater (biological treatment followed by disinfection) is recommended when unrestricted crops are irrigated [14–15]. In developing countries, most or the whole volume of wastewater produced in cities is treated prior to irrigation, while in low income countries wastewater treatment is not a priority, and thus untreated or partially treated wastewater or a mixture of treated and untreated wastewater is commonly used for agricultural purposes [12, 16]. In Mexico, China, India and Pakistan, for instance, large areas exist where untreated wastewater has been reused in agricultural irrigation for a considerable time [17]. The World Health Organization estimates that nearly 20 million hectares throughout the world are irrigated using untreated wastewater [18]. It is also reported that in some cities up to 80% of the vegetables locally consumed are produced using wastewater for irrigation [19]. The application of wastewater to soil, particularly untreated wastewater, followed by its infiltration poses a significant risk of pollution, not only to soil and crops but also to the surface and subterranean water sources surrounding the irrigated area [20–21].

Pollution by pathogenic agents is the main cause of concern regarding the application of treated/untreated wastewater to soil. Due to the variety of microorganisms entering the soil via the wastewater there is a high risk of enteric disease outbreaks for farmers and consumers

[22–23]. This chapter addresses the contamination of wastewater irrigated soils by helminths (intestinal worms) and pathogenic bacteria common in developing countries (where untreated wastewater is used to a greater extent), as well as the risk of outbreaks of parasitic diseases for both farmers and consumers in agricultural areas where untreated wastewater is reused. The occurrence of antibiotic resistance in indigenous organisms of soil and pathogens reaching soil via wastewater is gaining the attention of scientists and health organizations around the world [24–25], thus a review of what it is known and the research opportunities in this field are presented in the text. With regard to organic pollution, a current topic of interest is the entry to the soil and potential risks within crops of so–called "contaminants of emerging concern". These pollutants are substances that have not previously been considered as pollutants since they are part of everyday products; however, due to the subtle but harmful effects that these substances may cause in a variety of aquatic and terrestrial organisms, concerns have risen due to their continuous entry into the environment via wastewater [26]. A review on the presence of some organic contaminants of emerging concern, such as pharmaceutical substances, personal care products and industrial additives, in wastewater–irrigated agricultural soils is presented in this chapter along with some of the known potential effects caused to soil organisms, plants and consumers. Such effects have just begun to be elucidated, and only for some groups of contaminants of emerging concern [27–28], even though it is now known that up to 7 million commercially available chemicals are routinely disposed of in sewage after use [29]. In this regard, this chapter makes some suggestions regarding the next steps in the toxicity studies for this class of pollutants, such as testing the synergistic effects of mixtures of contaminants of emerging concern in soil organisms.

In spite of the variety and quantity of contaminants that soil regularly receives through wastewater irrigation, this ecosystem possesses self–purification processes that maintain homeostasis within the system. Such self–purification processes may either inactivate or reduce the population of pathogenic microorganisms reaching the soil via wastewater through predation by the indigenous microbiota within the soil [30–31], the production of antibiotics by some organisms in the rhizosphere [32] and by retention of microorganisms in the surface layers of the soil profile through physical and chemical processes. For organic pollutants, mechanisms such as photolysis and biodegradation promote the dissipation of contaminants in the soil, while adsorption onto the soil particles lead to the retention –and the potential confinement– of organics within the solid matrix [33]. In this chapter, current knowledge concerning the environmental fate of pathogen and organic contaminants of emerging concern in wastewater irrigated soils is discussed, highlighting the laboratory approaches that show the best results in simulation of the conditions in the field. Knowledge of the environmental fate of contaminants in irrigated soils is important in order to perform more accurate risk assessment studies on contamination of water sources, soil and crops in wastewater irrigated areas; furthermore, it provides information to policy makers to make proper legislation aimed at promoting environmentally responsible management of treated/untreated wastewater in agricultural irrigation.

Depuration of wastewater prior to its reuse is the most plausible option to prevent soil pollution by wastewater reuse. However, since wastewater represents a cheap source of water and fertilizer for farmers [34], it is necessary to consider the needs of users before planning schemes of wastewater treatment. The use of wastewater treatment systems aimed at removing carbon, nitrogen, phosphorous and minerals in wastewater leads to the reduction in quality of effluents as fertilizers, impacting crop yields and thus in the livelihood of farmers. In this sense, the use of advanced primary treatment systems could be a feasible option to: a) remove suspended solids, pathogens and heavy metals in wastewater without significantly impacting the content of nutrients in effluent; b) preserve the quality of agricultural soils to properly perform ecosystem services such as the production of food; and, c) fulfill the needs of farmers that use wastewater as a source of water and nutrients. Treating wastewater by these kinds of systems may be an opportunity to couple sanitation with reuse within a program of comprehensive management of wastewater, the recycling of nutrients and the use of soil as a food producer and purification system.

This chapter aims to describe what it is known and what it is unknown regarding the positive and negative impacts of the reuse of treated/untreated wastewater in agricultural irrigation. It will be shown in detail how this practice can benefit soil and farmers, while at the same time posing a risk of contamination to the ecosystem. Emphasis is given to the purification processes occurring in the soil and how soil manages the continuous entrance of pollutants via wastewater. Lastly, some perspectives for further studies on the presence and environmental fate of pollutants in wastewater irrigated soils are proposed.

IMPACTS OF WASTEWATER REUSE IN AGRICULTURE

The reuse of wastewater results in both beneficial and negative impacts on soil, some of which are explained in this section. The aim is to identify both and to understand their origins in order to assist scientists and policy makers to balance them and even to greater advantage of the benefits compared to the drawbacks in certain situations.

Benefits of Wastewater Reuse in Agriculture

Figure 1 summarizes the positive impacts of reusing wastewater in agricultural irrigation in all of its forms. The extent of the positive impacts depends on local conditions of the specific project.

Benefits in Crops

Since wastewater is produced constantly and thus is always available, it is possible to select a wider range of crops to be sown year–round, specifically those of high profitability which normally have higher and more stringent water demands in terms of quantity and timing. The consistent use of wastewater in irrigation may stabilize the content of nutrients in the soil, even when growing crops with high nutritional requirements; this is because the continuous withdrawal of nutrients by plants is compensated by the constant input of organic and mineral components into the soil via wastewater. Examples of how the reuse of wastewater has led to increases in crop yields in arid zones can be found worldwide. Studies conducted in Hubli–Dharwad, India, showed that irrigation with treated and untreated wastewater made it possible to produce vegetables during the dry season; yields and selling prices increased by 3–5 times compared to the kharif (monsoon) season [35]. In Pakistan, Ghana and Senegal the reliability and flexibility of wastewater supply allows rural and urban farmers to cultivate profitable crops in a shorter time, resulting in several harvests per year (3 to 6)

[36–37]. Treated/untreated wastewater is a source of organic matter and the same large diversity of nutrients contained in any formulated fertilizer. It is estimated that 1,000 m³ of municipal wastewater applied to one hectare can contribute 16–62 kg of organic nitrogen, 4–24 kg of phosphorus, 269 kg of potassium, 18–208 kg of calcium and 9–110 kg of magnesium each year [16]. Table 1 shows the contribution of water and nutrients that untreated wastewater make to several crops.

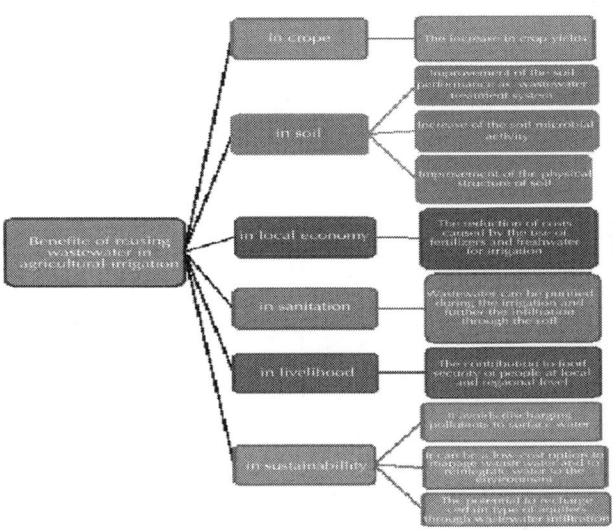

Figure 1: Beneficial impacts of reusing wastewater for agricultural irrigation. With information of references [14, 18, 21, 36, 37

Nitrogen is a plant macronutrient which can be found in the form of nitrate ions (N–NO$_3$), mostly in treated wastewater, or as ammoniacal nitrogen (N–NH$_4^+$) and organic nitrogen in untreated wastewater. The sum of all these forms is known as total nitrogen (TN). Most crops absorb nitrates to the greatest extent (85% of the nitrate contained in wastewater); whereas 50% of ammoniacal and less than 30% of organic nitrogen contained in wastewater can be assimilated as it is by plants. The remaining nitrogen is taken up by soil microorganisms and transformed into nitrates or volatilized as N$_2$. In wastewater irrigated soils; organic nitrogen is transformed into nitrates by soil microorganisms to a greater extent than that observed in non–irrigated agricultural soils [38]. Problems related to high inputs of nitrate ions are due to their high solubility in water, and thus their rapid percolation through the soil to the aquifer.

Table 1: Contribution of nutrients and sodium from untreated wastewater and water requirements of demandant crops

Crop	Water requirements (mm/year)	Nutrients and sodium contribution by WW (kg ha/year)					
		Ntotal	Ptotal	K	Ca	Mg	Na
Maize	673	108–418	27–162	13–465	121–1401	61–741	182–1226
Green tomatoes	653	104–405	26–157	13–451	118–1358	59–718	176–1188
Chili	601	96–373	24–144	12–415	108–1250	54–661	162–1094
Beans	370	59–229	7–89	7–255	67–770	33–407	100–673
Wheat	520	83–322	21–125	10–359	94–1082	47–572	140–946
Barley	516	83–320	21–124	10–356	93–1074	46–568	139–939
Alfalfa	1360	218–843	54–326	27–938	245–2829	122–843	367–2475
Marrow	364	58–226	15–87	7–251	66–757	33–400	98–662
Oats	353.6	57–219	14–85	7–244	64–735	32–389	95–644

[i] - Source: reference [16]

A significant quantity of nitrate leaching through soil subsequently becomes unavailable for plants; this does not necessarily represent a problem, as nitrate is continuously supplied to soil via wastewater. More important, the presence of nitrates in subterranean water is related to occurrence of methemoglobinemia disease in infants ingesting nitrate at levels higher than 45 mg/L via drinking water [39]. The quantity of nitrogen washed out from soil depends on the irrigation rate, the frequency of rain events, the type of crops sown and the characteristics of the soil [40]. The amount of nitrogen that can be applied to soil to produce minimal nitrate leaching rates depends on the demand of crops, which usually varies between 50 and 350 kg of nitrogen per hectare [40]. Such demand is within or slightly above the amount of nitrate supplied by treated wastewater. In this sense, the limited removal of nitrogen by wastewater treatment would not significantly affect the input of this macronutrient to agricultural soils.

Phosphorous is another plant macronutrient, which is very scarce in soil, at the point it needs to be added through the application of fertilizers. Due to its stability and low solubility, this nutrient can be accumulated in soil. Wastewater normally contains small amounts of phosphorous, so its use for irrigation is beneficial to plants and it does not impact negatively upon the environment, even if applied consistently for long periods of time [40–41]. The recycling of phosphorous and nitrogen in wastewater–irrigated soils is important because it allows closure the P cycle rather than its breakage. Breakage of the cycle occurs when phosphorous is removed from wastewater during treatment, becoming trapped in sludge and dumped to confinement sites or landfills. An advantage of the availability of phosphorus in wastewater is that it is partly bound to organic components and thus it cannot form complexes with iron or aluminum ions upon its entry to soils [16]. In contrast to phosphorous, potassium is contained in soil at high concentrations (around 3% of the lithosphere) but in chemical forms that impede its bioavailability. As a result it is necessary to add potassium to soils via fertilizers. Approximately 185 kg of potassium per hectare are required to cultivate some crops [16]. Sewage contains low concentrations of potassium, insufficient to cover the theoretical demand in most cases. Meeting the demand for potassium in irrigated soils will depend on the amount of wastewater supplied at each irrigation event, the wastewater quality and the frequency of irrigation. Fertilization with potassium has not resulted in adverse impacts to the environment [42]. Recycling nutrients by the reuse of wastewater promotes savings in energy, which would otherwise be consumed in the production of fertilizers [43]. In particular, the recycling of phosphorus is important since the world's phosphorus reserves are becoming scarce [44]. Fertilizing agricultural soils by the reuse of wastewater invariably leads to the increase of crop yields. An example of this can be found in Mezquital Valley, Mexico [45]; in this respect, Table 2 shows the differences in the agricultural production in croplands of Mezquital Valley when either untreated wastewater or groundwater is used for agricultural irrigation.

Table 2: Comparison of crop yields for some vegetables in plots where wastewater and groundwater are used for agricultural irrigation (Mezquital Valley, central Mexico)

Crop	Crop yield (tons/ha)		Increment (%)
	Untreated wastewater	Groundwater	
Corn	5.0	2.0	150
Barley	4.0	2.0	100
Tomato	35.0	18.0	94
Oats for forage	22.0	12.0	83
Alfalfa	120.0	70.0	71
Chili	12.0	7.0	70
Wheat	3.0	1.8	67

[i] - Source: references [16, 52]

The use of wastewater in Mezquital Valley has also contributed to changing the landscape of the zone, transforming barren soils into productive and green vibrant soils, as shown in Figure 2.

Figure 2: Comparison of untreated wastewater irrigated (right side) and rain–fed (left side) croplands from Mezquital Valley, central Mexico.

Benefits in Soil Quality

In order to define the improvements in soil quality produced by the application of treated/untreated wastewater it is necessary to establish the use of the irrigated soil. It is known that soil complies with five ecological functions: a) a medium for plant growth (including agriculture); b) a biodiversity pool and habitat for plants and (micro and micro) fauna; c) a carbon sink; d) a storage, filter and transforming medium for nutrients, pollutants and water; and, e) a landscaping and engineering medium. [46]. This chapter focuses on the functions of soil as a medium for plant growth as well as in its role as a transforming medium for nutrients and pollutants.

In addition to the continuous supply of nutrients to the soil, irrigation with treated/untreated wastewater confers significant improvements in soil quality. Favorable changes reported in irrigated soils comprise: a) an improvement in the physical structure of soil; b) an increase in soil microbial activity; and, c) the improvement of the soil performance as a wastewater treatment system.

Improvement of the Physical Structure of Soil: The physical structure of soil is defined as the arrangement of the solid particles and the size, shape and interconnection of pores and voids. Soil structure is closely related to its capacity to store and transport gases and water (and thus dissolved substances) [47]. Gas exchange between the soil and the atmosphere determines whether aerobic, anoxic or anaerobic conditions prevail within the soil. This in turn regulates the metabolism of soil microorganisms and impacts, inter alia, upon the nitrogen fixation, the transformation of soil organic matter and the degradation of pollutants. Additionally, the physical structure of soil affects the plant growth by influencing root distribution and thus the ability to take up water and nutrients [48]. Improvements in the physical structure of soil are related to the increase in both the stability of the soil aggregates and soil porosity. The enhancement of the physical structure of soil results in a rise in agronomic productivity, the augmentation of water infiltration through soil to the aquifer and a decrease in erodibility [49]. The hierarchical theory of aggregation proposes that microaggregates (particle size below 250 μm) in the soil are formed initially by the attachment of organic material to some inorganic components of soils (e.g. clay and hydroxides); in turn these microaggregates join together to form macroaggregates (particle size above 250 μm). Alternatively, macroaggregates can form around the particulate organic matter, while exudates produced by soil microorganisms serve as cementing agents, making micro and macroaggregates more stable [50]. Microaggregates can be also formed from bacterial colony clusters which use bacterial polysaccharide exudates to bind with clay particles. The clay particles act as a protective shell for clusters and macroaggregate formation continues as described above [51].

Since the formation of aggregates in the soil is related to the presence of organic matter, and in some cases microorganisms, it might be expected that the continuous supply of these two elements via wastewater would result in the increased formation and stability of soil aggregates and thus an improvement in the physical structure of soil. For example, the study referred in [52] establishes that increased soil microbial activity due to the augmentation of organic carbon content by the application of wastewater impacts positively upon the stability of soil aggregates. Furthermore, there are substances contained in wastewater other than organic matter and microorganisms that may contribute to the formation and stability of soil aggregates. Calcium and magnesium cations, which are abundant in wastewater, increase the formation of microaggregates through cationic bridging between clay and organic matter, resulting in aggregation. In arid soils and soils with low organic matter contents, insoluble calcium and magnesium carbonates can trigger the formation of soil micro and macroaggregates

[51]. Additionally, calcium can inhibit clay dispersion, and thus the breakup of aggregates, when sodium concentration increases in soil [53]. Dissolved organic matter in wastewater can form complexes with iron and aluminum in soil forming mobile organo-metallic compounds which can further precipitate and act as cores for microaggregates formation. Particulate organic matter (i.e. suspended solids in wastewater) may enhance the binding of microaggregates to subsequently form macroaggregates; for instance, extracellular polysaccharides of microorganisms in the surface of suspended solids can act as binding agents in the formation of macroaggregates [54]. In the case of phosphorous, the formation of insoluble aluminum and calcium phosphates in the soil can induce the formation of microaggregates and additionally it may act as a macroaggregate binding agent [55]. The entry of certain chemicals to the soil via wastewater increases the stability of soil aggregates. For example, hydrophobic substances (e.g. surfactants, lipids and hydrocarbons) decrease the wettability of aggregates by inducing water repellency, which in turn leads to increased cohesiveness and low decomposition rates of soil aggregates [51]. Agricultural activities in wastewater irrigated soils may also contribute to the improvement of the physical structure of soil. Previous studies have found that some crops (i.e. maize, alfalfa and leguminous plants) have beneficial effects on the conservation of the physical structure of soil. Aggregation of soil particles tends to increase when planting crops characterized by high density and long length of roots; this is because chemicals released by roots (i.e. mucilage) enhance the stability of soil aggregates in the rhizosphere by increasing the bond strength and decreasing the wetting rate [56]. According to the study reported in reference [57], roots of leguminous crops increase the aggregation of soil particles. Corn (*Zea mays*) residues (leaves and shoots) also increase aggregation of soil particles compared with other crops; this is attributable to the liberation of phenolic compounds from plant tissues, since phenols favor the agglutination of particles and prevent wetting [57–58]. Municipal and industrial wastewater may also be a source of phenolic compounds to soil through irrigation, producing similar effects to those of corn wastes [59]. The study referred to in [58] demonstrated that the stability of soil aggregates is high for continuous cultivation of alfalfa (*Medicago sativa*), while the opposite effect was observed for soybean. This is attributable to the low concentration of phenols in the latter [60]. Some studies have addressed

the changes in the physical structure of agricultural soils caused by long–term irrigation with wastewater. The results of these studies show a decrease in soil porosity caused either by occlusion of pores by the suspended solids contained in wastewater or by the augmentation of micropores (radius < 0.01 μm) in the soil matrix [61–62]. Depending on the method of water application during irrigation, an increase in the compaction of soil may be observed in the plot after an irrigation event [63]. Soils irrigated by flooding exhibit high compaction while water dropping effects (erosion) may be observed in soils irrigated by spraying. In any of both cases, wastewater irrigated soils exhibit large populations of earthworms which may assist in the formation and connection of pores within the soil matrix. Undoubtedly wastewater contains agents that improve the physical structure of soil. However, studies performed so far show contrasting results, either an increase in the soil microporosity or soil compaction. It is therefore necessary to carry out studies aimed at measuring changes in the physical structure of soil throughout several irrigation cycles and for longer periods (months or years); additionally, it is of interest to assess changes in the physical structure of soil at landscape level (piedmont or catena), as it may be useful for evaluating the horizontal displacement of soil particles and nutrients.

Increase of Soil Microbial Activity: Either due to the extra supply of organic carbon or because of the addition of microorganisms via wastewater, microbial activity in wastewater irrigated soils tends to be higher than that found in non–irrigated soils [64–65]. This increase in the microbial activity of the soil brings benefits to both agriculture and the development of flora and fauna in the soil ecosystem. According to the study reported in reference [66], the C/N ratio in soils irrigated with wastewater for long periods tends to decrease by up to 45%, which implies an improvement in the nutritional conditions for soil microorganisms. The authors report an increase in the population of copinotrofic and oligotrophic bacteria (234 and 217%, respectively), as well as in the populations of actinomycetes (234%) and fungi (206%) in soils irrigated with wastewater for 100 years compared with those populations found in non–irrigated soils. Rises in the metabolic activity of soil, measured as the production of ATP and enzymatic activity have been also reported [65-66]. According to reference [66], soil enzymatic activity remained unchanged 20 years after wastewater irrigation ceased. In contrast, the study referred to in [67] shows that elevated microbial activity in soils irrigated with treated wastewater decreases after few days without irrigation. Due to the augmentation of the populations of bacteria, actinomycetes and fungi in the irrigated soil, a rise in the rhizospheric activity is experienced, resulting in: a) the increase in the growth and development of plants; b) high rates in stabilization of organic matter entering the soil through wastewater; c) higher performance of the depuration of wastewater and degradation of the pollutants fixed in the soil in comparison with non–irrigated soils; and, d) the improvement in the formation and stability of soil aggregates. The latter may be explained by the role of polysaccharides exuded by bacteria as transient binding agents, which initialize aggregation of soil microaggregates [67]. The transformation of carbon and nitrogen by soil microorganisms supports the proliferation of soil (micro and macro) fauna which is essential for soil formation as well as for the development of plants. According to the work referred to in [68], the use of treated wastewater to irrigate an agricultural soil over 20 years has resulted in the improvement of the metabolic efficiency of soil microflora to transform carbonaceous and phosphorous substances into nutrients readily available to plants and macrofauna.

Soil biomass has proven to be capable of adsorbing a certain proportion of heavy metals contained in the wastewater. For instance, the study referred in [69] found biosorption rates for cadmium and nickel within the range of 5 to 55 mg/g of biomass in a soil that had been irrigated with wastewater for two decades. In that soil, the predominant bacteria after irrigation were *Enterobacteriaceae* and *Pseudomonas*.

The effect of wastewater irrigation on soil nitrogen fixing organisms has been little studied. An increase in soil nitrifying activity accompanied by a low rate of denitrification has been observed in wastewater irrigated forest soils [70], while in the study referred to in [71] a peak in N_2O production in a soil irrigated with treated wastewater was reported, followed by an immediate drop in gas production. So far, the metabolic processes performed by different soil microbial species in wastewater irrigated soils have been little explored. However, it is important to keep in mind the important role that soil microorganisms play in both the development of the soil and plants as well as in the purification of wastewater when planning agricultural systems based on the reuse of wastewater. Even when soil microbial populations show some kind of resilience to a wide variety of contaminants, some other chemicals can cause not only toxic effects to soil microorganisms but the proliferation of pathogenic organisms and the occurrence of antibiotic resistance within the agricultural soils.

Improvement of Soil Performance as a Wastewater Treatment System: As it is known, the application and infiltration of wastewater through soil results in its purification. In practice, specific wastewater treatment systems are based on soil infiltration, which have been demonstrated to improve water quality to levels obtained using tertiary treatment systems [72–73]. Purification of wastewater is one of the ecological functions of soil; through this mechanism, soil maintains, at least partially, the quality of surface and groundwater bodies. The extent at which this natural system works is highly variable, from almost nonexistent to very high, depending on local conditions and types of pollutants. Table 3 shows the extent to which pollutants in wastewater are removed by infiltration through the soil. The application of wastewater to soil reduces the content of pathogenic microorganisms by 6–7 log units for bacteria and 100% for helminths and other protozoa. Total organic carbon can be reduced by up to 90%, while levels of recalcitrant compounds in wastewater, such as phosphorus (20–90%), nitrogen (20–70%), and metals (70–95%) are also reduced dramatically. In sewage, organic phosphorus (5–50 mg/L) is biologically converted to phosphate; subsequently, in alkaline or calcareous soils, phosphate precipitates with calcium to form calcium phosphate and remains available for plants. In contrast, in acidic soils phosphate reacts with iron and aluminum oxides to form insoluble compounds, which are unavailable to plants. Sometimes soluble phosphate is initially immobilized by adsorption onto soil particles and then slowly returns to insoluble forms, allowing for further adsorption of mobile phosphate. This process is generally known as phosphate aging [72].

Table 3: Processes in soil that improve the quality of the wastewater, relative to selected parameters

Variable	Effect
Organic matter	Biodegradable material is reduced by more than 90%, while less readily biodegradable material is adsorbed and later biodegraded or volatilized.

Nitrogen	Nitrogen is removed from water at a level similar to tertiary treatment systems by transformation in soil as well as by assimilation by soil microorganisms and plants.
Phosphorus	Phosphorous is reduced to levels of 1 mg/L or less by assimilation by plants.
Microorganisms	Helminth eggs and protozoa are easily removed by straining in the soil surface; bacteria and viruses can also be adsorbed onto the soil particles and then desiccated or killed by indigenous soil microorganisms. The performance of these processes depends on the texture, physical structure and organic matter content of soil.
Heavy Metals	Heavy metals can be removed by the formation of complexes with soil organic matter, precipitation or methylation at efficiencies of 70–95%.
Toxic organic compounds	Most are retained in soil and then biodegraded at different rates.

Most of the organic compounds (natural and synthetic) in sewage are rapidly transformed in soil to stable, and in some cases non–toxic, organic compounds (e.g. humic and fulvic acids). Actually, soil biodegrades a greater amount and variety of organic pollutants than that reported for water streams. Wastewater application to soil under controlled conditions (e.g. limited irrigation rate and intermittent flooding) permits the biodegradation of hundreds of kilograms of carbonaceous substances per hectare per day, with no impact on the environment [72]. Total organic carbon levels in wastewater are dramatically reduced from levels of 80–200 mg/L to 1–5 mg/L in the infiltrated water [74]. Heavy metals can be removed from wastewater during soil infiltration and confined within the organic domain of the soil for several hundred years. Metals are retained in the surface layer of the soil either by complexation with soil organic matter or by precipitation at high pH values. Only a small fraction of metals infiltrates to lower layers of the soil profile and even less can be assimilated by crops. For instance, around 80–94% of cadmium, copper, nickel, and zinc can be removed in the first 5–15 cm of the soil profile, 5–15% is leached to lower layers and only 1–8% can be absorbed by grass [75]. A similar process occurs with fluorine [76]. This phytoremediation process is used to treat wastewater in planned natural treatment systems such as wetlands. However, it is necessary to be aware that some edible crops are able to take up heavy metals to a greater degree than grasses [77].

The capability of soil to act as a filter and transforming medium for wastewater pollutants can be observed in both long–term and newly wastewater irrigated soils [72,78]. The operation of this natural purification system is closely related to the physical and chemical properties of the soil and thus modifications in soil characteristics caused by irrigation with wastewater may either improve or worsen the performance of this natural wastewater treatment system. The increase in the soil organic matter content is the main factor resulting in an improvement in the removal of biological, organic and inorganic pollutants as wastewater leaches through the soil. This is because soil organic matter promotes the immobilization of pollutants either by adsorption or formation of complexes, while at the same time stimulating the proliferation of degrading microorganisms [78–79]. Regularly, heavy metals are fixed in the upper layers of the soil profile by complexation with organic matter [65], thus organic matter enrichment in wastewater irrigated soils results in greater retention of heavy metals by the solid matrix. Heavy metals cannot be biodegraded but they may be modified by soil microorganisms. Biological methylation of metals and metalloids, such as selenium, arsenic and mercury, has been reported in wastewater irrigated soils. It is expected that this process is elevated in wastewater irrigated soils, where microbial biomass occurs at higher levels than in non–irrigated soils. Methylation of heavy metals leads either to reduced toxicity or increased loss of metals in soil through volatilization [80–81]. Another process observed in long–term wastewater irrigated soils, related to those aforementioned, is the potential of soil microorganisms to develop resistance to the harmful effects caused by the presence of heavy metals in the solid matrix [69, 82]. Such resistance is similar to that developed to antibiotics and has been reported for cadmium, chromium, zinc and nickel in soils irrigated with wastewater over the long term [69, 83]. It is plausible that the expression of these resistances results in an increase of heavy metal methylation in the soil, which allows soil microorganisms to survive and to continue with those metabolic functions that increase agricultural productivity and purify wastewater. With regard to organic contaminants, the increase in the soil organic matter content produces, in most cases, an incremental boost in the adsorption of solutes onto soil particles. Studies referred to in [84] found increased adsorption of organic contaminants (i.e. pesticides, pharmaceuticals and estrogenic hormones) in long–term wastewater irrigated soils compared to rain–fed soils from the same agricultural area. Organic compounds displaying high hydrophobicity are adsorbed by soil not only faster and to a greater extent but with greater strength than is observed for semi–polar and polar compounds [85]. The increase in the hydrophobicity of soil due to the application of wastewater increases the capacity of such soils to strongly retain non–polar organic contaminants within the solid matrix. The increase in the adsorption of organic pollutants by soil results in an extended retention time in the solid matrix, encouraging biodegradation processes. Similar to the results reported for adsorption, higher rates of biodegradation of organic pollutants have been observed in treated/untreated wastewater irrigated soils compared to non–irrigated ones [86]. This may be caused, on the one hand, by the continuous supply of organic matter to the soil via wastewater, which can be used by soil microorganisms as co–substrate in the biodegradation of target organic pollutants, and on the other hand, by the prolonged exposure of soil organisms to pollutants. The latter case can be understood as the acclimation of the degrading organisms to the occurrence of organic pollutants in the soil followed in the short term by the acquisition of the capability for using organic contaminants as a carbon source. The increase in the soil organic matter content caused by wastewater irrigation has a positive impact not only in the adsorption of organic compounds but also on the retention by soil of wastewater–borne pathogens. This is due to the high affinity of the cell membranes to the organic domain of soil. The study referred to in [87] reports a higher adsorption of enteric bacteria *Escherichia coli* and the enteric protozoa *Giardia lamblia* in long–term wastewater irrigated soils compared with long–term groundwater irrigated soils from the same agricultural zone.

In general terms, an increase in pH values has been observed in agricultural soils irrigated with treated/untreated wastewater; although in less cases soil pH tended to decrease following the application of wastewater [81]. The first phenomenon is attributed to the continuous addition of salts (carbonates, calcium, magnesium, sodium) in wastewater. The second case is explained by the high mineralization rate of organic matter in the irrigated soil, which is highly dependent on the soil type, the climatic conditions of the site, and the quality of wastewater, among other reasons. The increase in soil pH, in combination with the continuous supply of organic matter, results in the buffering of soil pH, which prevents the drop of soil pH values during rain events (including acid rain). Stabilization of soil pH values also contributes to the retention of heavy metals in the surface layers of the soil by the formation of insoluble basic salts. Furthermore, basic values of soil pH can facilitate the adsorption of neutral and basic organic contaminants; as these compounds tends to be better adsorbed to neutral and basic soils than to acidic ones.

Since wastewater irrigation improves the physical structure of soil (i.e. increased formation and greater stability of aggregates), aerobic conditions may be maintained within the soil matrix; which in turns contributes to an increase in the aerobic biodegradation rate of organic pollutants. Additionally, an increase in the adsorption of pollutants can be achieved in better structured soils due to the increase in the specific surface area of soil particles. Moreover, higher biodegradation of the adsorbed contaminants can be expected as long as they remain available to microorganisms after adsorption. In irrigated soils where occlusion of the pores by the suspended solids in wastewater occurs, anoxic conditions may be achieved. Under such conditions, toxic species of heavy metals are chemically reduced into non–toxic species (e.g. Cr^{+6} into C^{+3} and As^{+5} into As^{+3}), then they may be immobilized by the formation of insoluble hydroxides. The extent to which wastewater irrigation contributes to the function of the soil as filter and degradation medium for pollutants is just beginning to be studied. The potential of soil to act as an efficient wastewater depuration system is a powerful argument to convince policy makers that agricultural irrigation with treated/untreated wastewater can be an appropriate strategy to simultaneously solve problems of water stress and low agricultural productivity with no negative impacts in the quality of water sources surrounding the irrigation site. This, of course, is achieved when all of the appropriate precautions to avoid contamination are taken at each site.

Negative Impacts of Wastewater Reuse in Agriculture

The main drawback of reusing treated/untreated wastewater in agriculture is the pollution of soil, the potential contamination of crops and water sources, and the inherent risk of harmful effects that contamination poses to the exposed organisms. Even when soil acts as an efficient living filter to remove, inactivate and transform the pollutants contained in wastewater, it is not fully effective at eliminating some of them. Moreover, as a result of the increasing industrial development, wastewater irrigated soils continuously receive newly synthesized substances, which may negatively impact the effectiveness of soil as a treatment system by poisoning the degrading microorganisms, destroying the physical structure of soil or damaging the natural cycles occurring within soil. The pollutants received by soil via wastewater may be different in developing and developed countries. Examples of this include pathogenic microbial agents. In developed countries most wastewater is treated prior to reuse and thus pathogens are not present in irrigation water, while in developing countries untreated wastewater is used in most of cases. Pathogens vary for different zones; for instance, the enteric protozoa *Giardia* is commonly found in wastewater of developing countries (Latin American and African countries), while the parasitic protozoa *Cryptosporidium* occurs in developed countries (United States and western European countries). Similar to microorganisms, some organic pollutants can be found in wastewater from developing countries and not in developed countries. Examples include some herbicides (e.g. DDT and atrazine) whose use is restricted in developed countries; on the other hand, nanomaterials and new–generation antibiotics, all of which are much more likely to occur in wastewater of developed countries. The determination of pollutants in soil initially requires specific sampling methods which take into consideration the heterogeneity of the soil matrix. In addition, specialized extraction techniques able to efficiently isolate analytes (or microorganisms) from soil are necessary prior to analysis. Specialized analytical methods have been developed and validated for the determination of trace contaminants and microorganisms in soil. However, in most cases, these methods are time–consuming, expensive and require the use of specialized reagents and personnel. It is therefore necessary to continue research towards the development of simpler and environmentally–friendly analytical techniques. Determining the occurrence and concentration of contaminants in soil is a task that requires a significant effort; however, this is only a part of the job. The study and understanding of the environmental fate of contaminants in soil is also a priority task to accomplish truly useful environmental risk assessment studies comprising soil, water sources, crops, farmers and consumers. Knowing the environmental fate of contaminants in the soil is necessary to understand the potentialities and limitations of each soil as a natural purification system of wastewater and an effective tool to define the capacity of each site to support wastewater irrigation in agriculture. Since soil is a complex and heterogeneous matrix, the fate of contaminants can vary significantly from one site to another. In this sense, it is worth defining which parameters are determinant in the fate of contaminants within soil and, on the basis of this knowledge, elucidating the fate of contaminants in other sites using mathematical tools to achieve such extrapolations. In this section, attention will be focused on pathogenic microbial agents, heavy metals and organic pollutants contained in municipal wastewater. The occurrence of such pollutants in wastewater–irrigated soil as well as their environmental fate in soil is addressed; additionally the most significant effects of these contaminants will be treated in some detail. Lastly, perspectives for further studies on the occurrence and fate of the studied pollutants in soil are presented.

Soil Pollution by Pathogenic Microbial Agents

Contamination of soil and crops by pathogenic agents is the effect of wastewater reuse in agriculture that receives most attention from environmentalists and scientists. Municipal wastewater contains a huge quantity and variety of bacteria, protozoa and viruses passed from human and animal feces and urine; therefore this water is a vector for intestinal infections (although some other diseases can spread from the environment via wastewater). Exposure may be direct through contact or ingestion of wastewater and soil, or indirect through contact with sick people or by ingestion of polluted crops, meat or milk. There are four groups at risk: a) farmers and their families, b) crop handlers, c) product consumers and d) people living nearby to irrigated fields. For any of these groups children and elderly are the most vulnerable, especially when they are undernourished. The most affected group is agricultural workers due to their high exposure to wastewater and contaminated soils [18]. Table 4shows the risk of infection of water–borne diseases for vulnerable groups in irrigated areas using treated/untreated wastewater.

Effects Caused by Microbial Pollution in Soil: Several diarrheal outbreaks have been associated with the use of wastewater to irrigate [18, 88]; however, since this occurs in places where sanitation, hygiene practices and drinking water are of low quality it is always difficult to define their specific contribution to the total diseases burden. Cholera, caused by the bacterium *Vibrio cholera*, is one infection closely linked to wastewater irrigation in poor countries. Other intestinal diseases related to the use of wastewater to irrigate are traveler's diarrhea caused by *Escherichia coli*, shigellosis caused by *Shigella* spp., gastric ulcers caused by *Helicobacter pylori*, giardiasis caused by the parasitic protozoan *Giardia intestinalis* and amebiasis caused by *Entamoeba histolytica*. Additionally, viral enteritis (caused by rotaviruses) and Hepatitis A are the most reported viral infections caused by consumption of polluted vegetables [89]. Some studies [90] report skin diseases, such as dermatitis (eczema), in farmers that come into contact with untreated wastewater and wastewater irrigated soil. Nail problems in farmers, such as *koilonychias* (spoon–formed nails), have also been reported as related to the presence of fungi in wastewater irrigated soils [91]. Health and growth problems have been observed in cattle that consume forage produced by wastewater irrigation. Furthermore, in low income areas where water is scarce, cattle are not only fed with fodder grown using wastewater but also they are allowed to drink the wastewater used for irrigation. Some protozoa can survive in the surface layers of soil or even in aerial parts of crops; animals can be infected after eating these crops, although this is a remote way of transmission. There is some evidence indicating that beef tapeworm (*Taenia saginata*) can be transmitted from livestock fed with wastewater–irrigated forage to meat consumers. Furthermore strong evidence indicates that cattle grazing on fields freshly irrigated with raw wastewater or drinking from raw wastewater canals or ponds can become heavily infected by*Cysticercus bovis*, the early stage of the *Taenia saginata* life cycle [88].

Microbial Agents in Wastewater Irrigated Soils: The study of microbial contamination by the use of treated/untreated wastewater in agricultural irrigation is focused in the pollution of crops rather than the soils receiving wastewater. This is because, on the one hand, a greater number of people are exposed to pathogenic microorganisms through consumption of contaminated crops, meat and milk than by direct contact with irrigated soils, and on the other hand, the difficulties in the analysis of microorganisms in soil; for instance, the inherent problems of extracting microorganisms from such a complex matrix as the soil. Studies in the Mezquital Valley, Central Mexico, found the occurrence of fecal contamination indicators (*Escherichia coli*). *Giardia lamblia* cysts and helminth eggs (*Ascaris lumbricoides*) at different depths of long–term wastewater irrigated soils. Results shown in Figure 3evidence the accumulation of the three microorganisms in the first few centimeters of the soil profile, indicating that infectious agents are removed from wastewater at the beginning of percolation through soil; such removal can be achieved by several physical and chemical phenomena. In this study, the content of pathogenic microorganisms in soils with different time under irrigation was also evaluated. Results showed that the accumulation of microorganisms in the tested soils is not related to the time under irrigation, suggesting that soils have mechanisms to inactivate and/or destroy these microorganisms after irrigation.

Table 4: Summary of health risk associated with the use of wastewater in agriculture

Group exposed	Helminth infections	Bacterial/viral infections	Protozoan infections
Consumers	Significant risk ofAscaris infection for both adults and children consuming vegetables contaminated with helminth ova.	Cholera, typhoid and shigellosisoutbreaks reported due to the consumption of polluted crops. Seropositive responses for Helicobacter pylori in crop consumers. Increase in risks of suffering non–specific diarrhea when concentration of thermotolerant bacteria in wastewater used for irrigation exceeds 104CFU/100 mL.	Evidence of parasitic protozoa found on the surface of wastewater– irrigated vegetables, but no direct evidence of disease transmission.

| Farm workers and their families | Significant risks of Ascaris infection for both adults and children in contact with untreated wastewater and irrigated soils. Risk remains, especially for children, when wastewater presents more than 1 nematode egg per litre. Increased risk of hookworm infection in farmers. | Increased risk of diarrheal diseases for children in contact with wastewater when it exceeds 104CFU/100 mL for thermotolerant coliforms. Elevated risk of Salmonella infection in children exposed to untreated wastewater and wastewater irrigated soil. Elevated seropositive responses to norovirus in adults exposed to partially treated wastewater and wastewater irrigated soil. | Risk of Giardia intestinalis infection insignificant for contact with both treated/ untreated wastewater and soil. Increased risk of amoebiasis observed due to contact with untreated wastewater and wastewater irrigated soil. |
| Nearby communities | High risk of infections when flood and furrow irrigation is used. Ascaris transmission not studied for sprinkler irrigation. | Sprinkler irrigation with untreated wastewater and high aerosol exposure associated with Increased rates of bacterial infections due to the use of partially treated wastewater (104–105 CFU/100 mL or less). No risks of viral infection associated with sprinkler irrigation. | No data of protozoan infections transmission during irrigation with wastewater. |

As mentioned above, different types of microorganisms can be found in wastewater irrigated soils depending on the zone where reuse is taking place. For example, the study referred in [92] showed a higher prevalence of *Cryptosparidium* spp. compared with *Giardia* spp. in wastewater irrigated and manure amended soils of dairy farms in southeastern New York. *Cryptosporidium* is a protozoan commonly found in developed countries, while different species of *Giardia* are widespread in developing countries. In this respect, the study referred in [93] found the occurrence of *Ascaris lumbricoides*, hookworm and *Trichiuris trichiura* in 69% of the soil samples taken in an untreated wastewater irrigated area in West Bengal, India.

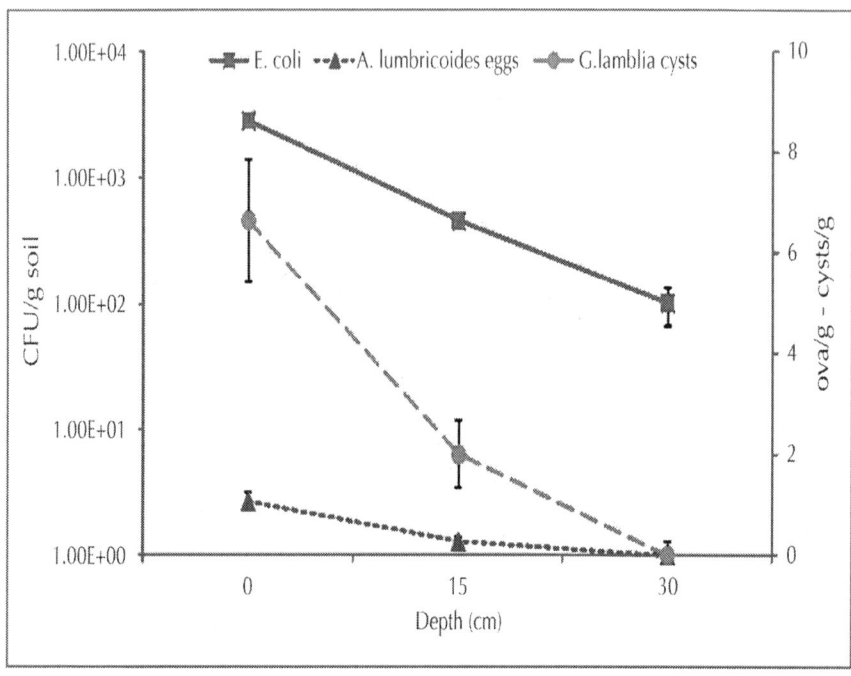

Figure 3: Abundance of three pathogenic microorganisms in a long–term wastewater irrigated soil at different depths.

The entry of antibiotic–resistant pathogens (ARPs) and antibiotic resistance genes (ARGs) into the soil via wastewater is an emerging issue. Since municipal wastewater contains both sub–therapeutic amounts of antibiotics, ARPs and ARGs –which occur to a greater extent when sewer systems combine municipal and hospital wastewater– [94], these substances can reach the soil, modifying the dynamic of soil microbial populations. Antibiotic resistance may occur naturally in the soil, and to a greater extent in the rhizosphere, which functions as a hotspot for both antibiotic–resistant bacteria and ARGs [32]. Previous studies have found the presence of opportunistic pathogens (*Stenotrophomonas maltophilia*, responsible for respiratory tract infections and endocarditis) in the rhizosphere of *Brassicacaea* type plants [95]. The transfer of ARGs from these opportunistic bacteria to human pathogens reaching the soil through wastewater has not yet been demonstrated. The ARPs reaching the soil through wastewater may survive on the soil surface and, if conditions are appropriate, reproduce or migrate to surface and groundwater sources. ARGs may be mobilized into aquifers by infiltration of wastewater or into surface water sources by runoff. So far a relationship between the presence of traces of antibiotics in wastewater and the occurrence of antibiotic resistance in the irrigated soil has not been categorically established. In previous studies the incidence of two sulfonamide resistance genes (*sul1* and *sul2*) was determined in the Mexico City wastewater, agricultural soils irrigated with such wastewater over different time periods and rain–fed soils [96]. The authors found the presence of ARGs in the three analyzed matrices; the concentration of resistance genes was 150 to 1500 times higher in irrigated soils than in non–irrigated ones. The occurrence of ARGs was positively related to the time under irrigation, with a higher content of resistance genes occurring in *Enterococci* bacteria living in soils irrigated for longer periods of time [96]. Such behavior may indicate that prolonged irrigation with wastewater promotes both the proliferation of indigenous ARPs in soil, due to the high and constant supply of nutrients via wastewater, and the increase in the assimilation of resistance genes due to the higher biomass content in old wastewater irrigated soils. Conversely, studies reported in [97] found that the abundance of isolates resistant to tetracycline, ciprofloxacin, sulfonamides and erythromycin were identical in wastewater irrigated soils and freshwater irrigated soils despite the high load of ARGs and ARPs in the wastewater used for irrigation. In this regard, the study in reference [98] found, by comparing the resistome of soils irrigated either with wastewater or groundwater, that *Entetococci* bacteria in freshwater irrigated soil were highly resistant to a greater number of antibiotics (erythromycin, tylosin, tetracycline, and ciprofloxacin) than long–term wastewater irrigated soil, which showed resistance to lincomycin and daptomycin. Furthermore, no differences were found in the content of ARPs when wastewater and freshwater irrigated soils were compared, suggesting that ARPs rarely survive after they enter soil via wastewater. Even though it seems unlikely that development of antibiotic resistance to human pathogens in wastewater irrigated soil is related to the input of antibiotics and resistant organisms via wastewater, it is worth, as a next step, studying the exchangeable genetic material (e.g. plasmids), since such material can be assimilated by soil microorganisms, inducing antibiotic resistance. Many questions remain about the mechanisms leading to the transference of this type of genetic material [99].

Microbial Pollution in Crops: Crops are polluted by direct contact with wastewater during irrigation. Pollution of the edible parts of plants depends not only on the quality of water, but also on the quantity applied to soil, the irrigation method and the type of crop. For example, zucchini when spray–irrigated with wastewater accumulate higher levels of pathogens on their surface than other crops. Zucchini have a hairy and sticky cover and grows close to the ground, which favors the attachment of pathogens. Microbial contamination of crops can occur not only as a result of wastewater irrigation but also during washing, packing, transportation and marketing. These problems are frequently not addressed, giving the impression that irrigation is the only source of microbial pollution [100]. In a previous study, referred in [101], it was found that less microbial pollution of crops is caused if irrigation is performed by subsurface dripping than through sprinklers, furrows or flooding. Moreover, the study reported in [102] showed that subsurface irrigation does not pollute crops even when using wastewater with $6–7 \times 10^5$ CFU/100 mL of fecal coliforms and 225 helminth ova/L. Microbial pollution of crops also depends on the type of crop. Fruits from trees are rarely polluted when irrigation is not provided using sprinklers (this is not a common procedure used to apply wastewater since sprinkler heads tend to become clogged). Fruits grow far from the watering sites when furrow and flood methods are used. The microbial contamination of crops in wastewater irrigation systems is closely related to the survival of microorganisms. Table 5 shows the survival times of some pathogens in agricultural soils and crops irrigated with wastewater.

Table 5: Survival of selected pathogens in soil and crops irrigated with wastewater

Pathogen	Survival time (days)	
	Soil	Crops

Ascaris lumbricodes eggs	180	30
Salmonella spp.	80	25
Fecal coliforms	<70, but usually <20	<30, but usually <15
Vibrio cholera	<20, but usually <10	<5, but usually <2
Entamoeba histolytica	<20	<10
Trichuris trichiura eggs	>180	<60, but usually <30
Taenia saginata eggs	>180	<60, but usually <30
Enterovirus	<40	<20

[i] - Source: references [105–106]

Both pathogenic and non–pathogenic microorganisms display differences in their survival in soil and crops. For instance, the non–pathogenic fecal coliform indicator *E. coli* can survive in soil for nearly a month, while the pathogenic strain of *E. coli* O157:H7 survives at most for 14 days in spinach leaves [103]. It is known in some detail that survival of pathogenic bacteria can increase by internalization within the plant tissues [104]. Previous studies indicate that *E. coli* can translocate from soil to leaves of lettuce through the root system [107]. In contrast, the results reported in reference [108] indicate that translocation of pathogenic bacteria to the edible parts of crops via the root system is quite unlikely. It is more likely that pathogens enter to the edible parts of crops through wounds in vegetal tissues [109]. Wounded tissues have been demonstrated allow the entrance of *Salmonella* and *E. coli* to lettuce and tomato plants [110–111]. Similarly, it is reported that *E. coli* can use the stomatal cavities in leaves to enter the internal structure of lettuce [115]. The pathway of this kind of entry is still unknown. Once inside the plant tissues, pathogen survival rates improve since they can use cellulose as their main source of carbon. Protozoa are larger in size than bacteria and thus they cannot access the internal parts of the plants; however, these pathogenic organisms can adhere to the surface of edible plants and remain there by the excretion of polymers which facilitate adhesion. Table 6 shows some examples of the occurrence of protozoa in crops irrigated with treated/untreated wastewater.

Table 6: Occurrence of some pathogen protozoa on the surface of crops irrigated using treated/untreated wastewater

Pathogen	Crop	Occurrence	Reference
Giardia lamblia	Potatoes	5.1 cysts/kg	Crops irrigated using untreated wastewater in Marrakesh. [112]
	Coriander	254 cysts/kg	
	Mint	96 cysts/kg	
	Carrots	155 cysts/kg	
	Radish	59.1 cysts/kg	
Ascaris lumbricoides	Potatoes	0.18 eggs/kg	
	Turnip	0.27 eggs/kg	
	Coriander	2.7 eggs/kg	
	Mint	4.63 eggs/kg	
	Carrots	0.7 eggs/kg	
	Radish	1.64 eggs/kg	
Enterobius vermicularis	Lettuce	10–40 cysts/kg	Crops irrigated using treated and untreated wastewater in Kahramanmaras, Turkey. [113]
	Parsley	10–60 cysts/kg	
	Cress	10–20 cysts/kg	
	Spinach	1–3 cysts/kg	
Entamoeba hystolitica	Lettuce	10–50 cysts/kg	
	Parsley	10–50 cysts/kg	
Giardia lamblia	Lettuce	10–20 cysts/kg	
Ascaris lumbricoides	Lettuce	10–30 eggs/kg	
	Parsley	10–30 eggs/kg	

Trichuris trichiura	Spinach Pudina Coriander	3.3% of the analyzed samples. 3.1% of the analyzed samples. 5% of the analyzed samples.	Crops grown in soils irrigated with raw wastewater in West Bengal, India. [93]
Hookworm	Lettuce Parsley Spinach Pudina Celery Coriander	9.4% of the analyzed samples. 3.3% of the analyzed samples. 6.7% of the analyzed samples. 9.4% of the analyzed samples. 3.6% of the analyzed samples. 5% of the analyzed samples.	
Ascaris lumbricoides	Lettuce Parsley Spinach Pudina Celery Coriander	43.8% of the analyzed samples 23.3% of the analyzed samples. 36.7% of the analyzed samples. 50% of the analyzed samples. 25% of the analyzed samples. 35% of the analyzed samples.	
Helminth eggs	Leafy vegetables Cauliflower	100 eggs/kg	Vegetables irrigated with untreated wastewater in Faisalabad, Pakistan. [114]

Fate of Pathogenic Microorganisms in Soil: Upon their arrival to irrigated soils, microorganisms can either survive or be inactivated / killed by the physical and chemical processes naturally occurring in soil as well as by predation by indigenous soil organisms. Given the case that these microorganisms can survive in the soil, they may subsequently colonize soil particles, infiltrate the soil to the aquifer or migrate through across the landscape by runoff. Processes affecting the environmental fate of the pathogenic microorganisms in soil are shown in Figure 4. Previous experiments have demonstrated that some microorganisms can vertically and/or horizontally mobilize through the soil, travelling long distances from the initial point of contamination [116]. Bacterial migration in soil has been reported up to 830 meters, while for viruses such displacement is significantly lower, i.e. up to 408 m [117–118]. Survival of pathogens is related with their environmental fate since the longer the lifetime of the microorganisms the larger the distance they can travel. As indicated in Table 5, bacteria can survive for long periods compared to viruses, and thus bacteria can be transported farther. Climatic conditions also impact upon pathogen transportation; for instance, in frozen soils pathogens can survive longer and thus they can be transported farther than in tropical and desert soils [119]. Microorganisms can be more easily displaced through coarse textured soils than fine textured ones. The study referred to in [118] found greater mobilization of coliforms in sand–gravel soil than in fine sand. In fact, in coarse sandy soils, the vertical movement of microorganisms can be as rapid as that observed for inorganic tracers. In this regard, the results reported in reference [120] evidence that infiltration of streptomycin–resistant *E. coli* can be compared with that of the chloride tracer in undisturbed soil columns, even when different soil textures are compared. Since the transportation of microorganisms is similar to that observed for tracers, the physical structure of soil is the determinant factor in them reaching the aquifer; therefore, a greater occurrence and interconnection of pores within the solid matrix may result in efficient infiltration of water and thus bacteria. Studies on the movement of pathogens in the field confirm the rapid movement of pathogenic bacteria observed in laboratory tests. These studies also found a high concentration of bacteria and viruses in groundwater [121]. In addition to the higher quantity and interconnection of pores, the increased transport of bacteria and viruses through the soil can be explained by the presence of preferential paths within the soil matrix.

Such preferential paths are referred to cracks, fractures, worm holes and channels formed by plant roots or fauna in the soil. Studies reported in reference [122] show that larger microorganisms (*E. coli*) can mobilize deeper into soil than smaller coliphages. Moreover, the study referred to in [123] confirms that bacterial cells smaller than 1 μm in diameter are more rapidly transported through soil than larger organisms.

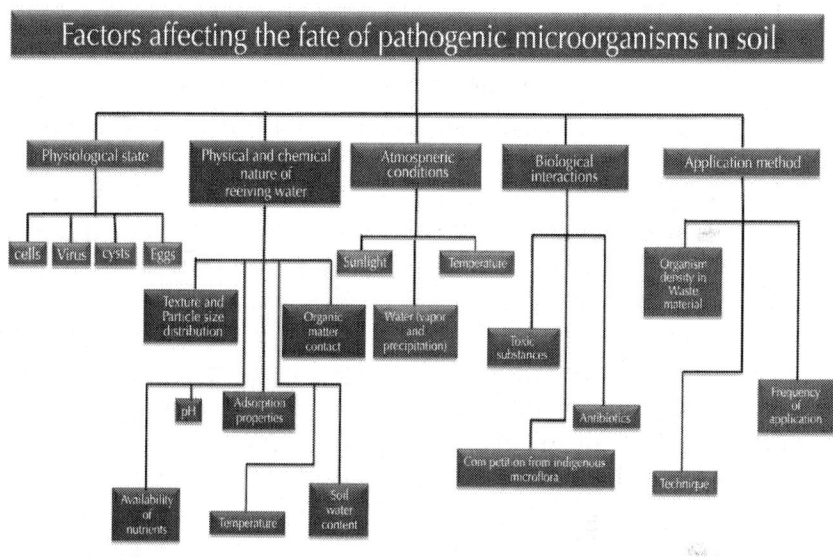

Figure 4: Factors affecting the environmental fate of pathogenic microorganisms in wastewater irrigated agricultural soils.

Source: modified from [121]

The chemical properties of soil can also impact upon the vertical and horizontal transport of microorganisms. The mineral composition of soil can favor adhesion of microbial cells, eggs or cysts onto soil particles. Several types of bacterial cells have been shown to strongly adhere to the mineral domain of soil and aquifer material [124]; once adhered, bacteria can replicate and form biofilms on the surface of soil particles. In wastewater irrigated soils, the accumulated organic matter as well as the continuous input of dissolved organic matter via wastewater may enhance the proliferation of bacteria. With regard to parasites, the study referenced in [87] found that *Ascairs lumbricoides* eggs and *Giardia lamblia* cysts adhere to the mineral fraction of wastewater irrigated soils more rapidly and more strongly than to the organic domain. In the case of *Ascaris* eggs, adhesion occurs with the silica in sand particles. In contrast to protozoan eggs, studies referred to in [125] suggest that adhesion or adsorption of protozoan cysts may be related to soil organic matter rather than the mineral fraction of soil. This has been attributed to the hydrophobic nature of the cysts walls. According to the findings reported in [126] detachment of bacteria from soil particles is effected by the composition of the irrigation water. In that study, *Pseudomonas* sp. showed enhanced transport when distilled water was used for detachment in column experiments, compared with 0.01 M NaCl. Such results suggest that clean water can efficiently wash off the polysaccharides excreted by bacterial cells which act as an adhesive between soil particles and bacteria. The opposite effect has been observed for Ascaris eggs. When soil is washed with NaOCl, eggs are effectively detached from soil particles; this is because sodium hypochlorite can destroy the albuminose layer that coats the surface of helminth eggs and which anchors with the soil particles [127]. According to the established in [87], the environmental relevance of studying the impact of this salt on the detachment of eggs from soil relies on the fact that NaOCl can be found in reclaimed water, as it is commonly used for disinfection of effluents.

Once microorganisms are retained by soil, either by adsorption/adhesion or straining, they can be inactivated or eliminated by desiccation. This phenomenon is particularly important in arid areas where high levels of solar radiation are reported. The environmental fate of microorganisms in soil also depends on the native microorganisms living in the solid matrix. Predators of wastewater-borne pathogenic bacteria in soil include Streptomycetes, Myxobacteria, *Bdellovibrio* and nematodes [121]. The presence of plants may affect the persistence and movement of microorganisms in soils. On the one hand, pathogen can found favorable conditions for survival in the rhizosphere due to the high content of nutrients in this zone; and on the other hand, native bacteria in rhizosphere can be natural predator of those pathogens, while roots may excrete antibiotics that inhibit or kill pathogenic microorganisms.

Soil Pollution by Heavy Metals

Given that most agricultural wastewater irrigation is performed using municipal wastewater, which contains negligible amounts of heavy metals [11], the occurrence of these elements in wastewater irrigated soils is usually significantly lower than the maximum permissible concentrations established by international regulations. However, there are some cases where care should be taken when reusing wastewater in irrigation, e.g. close to tanneries, metal processing or mining areas [91]. Different levels of risk are perceived for the different heavy metals. While some of them are nutrients for plants at trace concentrations, others have been shown to produce harmful effects on exposed organisms, or are absorbed by plants and accumulated through the food web. Table 7 presents the risks that are incurred by the presence of some heavy metals in soil.

Table 7: Heavy metal risk characteristics during irrigation

Risks characteristics	Metal
Low risk	Mn, Fe, Zn, Cu, Se, Sb
High risk	Cr, As, Pb, Hg, Ni, Al, Cd
Essential micronutrient to plants	Cu, Fe, Mn, Mo, Zn, Ni
Beneficial for some crops	Co, Na, Si
Can accumulate in crops to levels that are toxic for consumers	Cd, Cu, Mo
No human toxicological threshold established for wastewater intended for irrigation	Hg
Relatively high threshold for wastewater used in irrigation	Cu, Fe, Mn, Zn
Low absorption by plants	Co, Cu, Mn, Zn

[i] - Source: with information from [75]

Cadmium is the metal with the highest associated risk. It is toxic to humans and animals in doses much lower than those that visibly affect plants; furthermore crop uptake (which is notably high in acidic soils) can increase the dose consumed by organisms and in turn accumulation in animal tissue. Absorbed cadmium in animals is stored in kidney and liver, although meat and milk products have shown to be little affected by cadmium accumulation [75]. There is a relatively good knowledge to allow the setting of limits regarding the acceptable amount of heavy metals contained in wastewater used to irrigate. In the study referred to in [128], numerical calculation of the limits for the maximum tolerable pollutant concentration in wastewater irrigated soils was carried out (health–based targets). This was based on the acceptable daily human intake (ADI) for selected heavy metals and the amount that can be "permitted" to accumulate in soil before harmful effects occur in consumers of crops (Table 8). This analysis assumed: a) only two exposure routes (wastewater → soil → plant → human; and, wastewater → crop → human); b) a global diet in which the daily intake of grains/cereals, vegetables, root/tuber crops and fruit accounts for ~75% of daily adult food consumption; c) a body mass for adults of 60 kg; d) all of the food grain, vegetables, root/tuber crops and fruits are obtained from land irrigated with wastewater; and, e) a total daily intake of pollutants by this consumption path of 50% of the ADI (the remaining 50% of the ADI was attributed to background exposure). Table 8shows the inputs of heavy metals by wastewater to irrigated soils, assuming an application of treated wastewater of approximately 1.2 m/year, which is roughly the amount of water required to produce a crop cycle in an arid zone.

Table 8: Maximum tolerable concentration of heavy metals in wastewater irrigated soils

Element	Maximum input by wastewater (kg/ha/year)	Maximum tolerable concentration (mg/kg)
Arsenic	0.6–12	9
Cadmium	0.06–0.24	7
Chromium	1.2–60	3200
Lead	1.2–60	150
Mercury	0.12–0.12	5
Nickel	0.24–12	850
Selenium	0.24–0.6	140
Silver	1.2	3

[i] - Source: reference [128]

Health effects associated with the use of water heavily contaminated with industrial discharges for agricultural irrigation have been reported. In Japan, itai–itai disease, a bone and kidney disorder associated with chronic cadmium poisoning, occurred in areas where rice paddies were irrigated with water from the contaminated Jinzu River [129]. In some parts of China, the use of industrial wastewater for irrigation was associated with a 36% increase in hepatomegaly (enlarged liver) and 100% increase in both cancer and congenital malformation rates [130].

With regard to the occurrence of heavy metals in agricultural soils irrigated using wastewater, the study referred in [131] presents an inventory of sources of some heavy metals (zinc, copper, nickel, lead, chromium and cadmium) in agricultural soils of England and Wales. Results showed that the greatest contribution of heavy metals in those soils comes from the application of sludge from wastewater treatment plants, while irrigation appeared to be of little importance as a source of heavy metals in soils. According to this investigation, which followed the rates of deposition of heavy metals in the studied soils, the time required for metal concentrations to reach maximum values permitted by international regulations is 80 years for zinc and at least 1256 years for cadmium. In this respect, study referred to in [132] showed that concentrations of heavy metals in long–term untreated–wastewater irrigated soils in central Mexico were 10 times lower than the limits set by the Danish regulations; moreover, the authors estimated that another century of irrigation is necessary to exceed these values. In most cases, metals have little impact on aquifers. According to the results reported in [133] the most toxic metals to humans –cadmium, lead, and mercury– were absent in groundwater at five sites in the United States after 30–40 years of applying secondary and primary effluents at rates between 0.8 m/year and 8.6 m/year to different crops. The reason given was that the pH values greater than 6.5 in soil and wastewater resulted in the precipitation of the entire amount of metals. Metals are normally bonded into the organic matter through the formation of organo–metallic complexes, which are not bioavailable to plants. The addition of lime and wastewater to soil assists the precipitation of metals, while the addition of chemical fertilizers has the opposite effect, since over the long term they tend to lower the soil pH and thus solubilize metals.

In contrast, agricultural soils have been reported in which the concentration of heavy metals, such as cadmium and zinc, are close to reaching the maximum levels set out in international regulations. In these cases, the factors leading to an exacerbated soil contamination and thus increased risk of groundwater and crop pollution are: a) sandy soil texture; b) acidic to neutral soil pH; c) low organic matter content; and/or, d) the use of industrial wastewater for agricultural irrigation [134–135]. In such cases, the cessation of agricultural irrigation with wastewater is recommended, together with allowing the recovery of soil through remediation techniques such as phytoremediation.

Soil Pollution by Organic Compounds

Pollution of soil by organic substances has been a matter of concern to scientists and organizations regulating the quality of soil, water sources and food for several decades. An extensive body of work exists addressing the degradation of soil by conventional organic pollutants (e.g. pesticides, polyaromatic hydrocarbons, organochlorides, paraffin, organic solvents, etc.). However, in sites where treated/untreated wastewater is disposed of by agricultural irrigation one can find organic substances different to those commonly studied and reported in literature treating oil spills, mining zones or soil polluted by industrial wastewater. Most of the dissolved and particulate organic matter contained in municipal wastewater is produced by the degradation of human and animal excreta, hence organic matter in wastewater is composed mainly by saccharides, lipids, amino acids and proteins; however, a tiny fraction of the organic material in wastewater originates from chemicals contained in everyday consumer products used and disposed of via sewage by people in urban and rural areas. According to [24], thousands of organic compounds are contained in municipal wastewater at trace levels and there is a lack of knowledge regarding the effects that such substances may cause to the exposed organisms, either by themselves or in combination with other compounds or groups of organic compounds. This group of chemicals is referred as "organic pollutants of emerging concern" (OPECs) [136]; though they should actually be listed as priority pollutants in cases where wastewater is used to irrigate crops, since these contaminants are in contact with soils, crops and water sources near the irrigation area [137]. Over the last three decades, significant work in the field of analytical chemistry has been carried out in order to extract, isolate and quantify some of these pollutants in wastewater and soils. Frequently found OPECs in such complex matrices are pharmaceutically active compounds (PACs) and their metabolites, personal care products (e.g. disinfectants, fragrances, insect repellents, sunscreens, etc.), sweeteners, stimulants (e.g. caffeine and psychoactive drugs), detergents and their metabolites, plasticizers and industrial additives (e.g. additives in gasoline) [137]. Almost all of the studies addressing the removal of OPECs in wastewater treatment plants report that most of these substances are partially degraded/removed in primary and secondary treatment systems –and some pollutants are only partially removed even in tertiary treatment systems– [138]. Because of this, OPECs occur in irrigated soils if either treated or untreated wastewater is used in irrigation. Effluents of wastewater treatment plants contain a small fraction of the parent substance as well as the by–products generated during treatment. However, some of the compounds may be retained and concentrated in the sludge produced during wastewater treatment and reach the environment via the use of sludge (or biosolids) as soil amendments in agriculture. Due to continuous industrial development, the number of organic substances contained in wastewater is constantly increasing; in fact, most of these substances are not tested before they are released onto the market, and therefore their potential risks or the side effects they cause in non–target organisms in soils or water bodies is yet unknown.

Effects Caused by Domestic Wastewater–Related Organic Pollutants in Irrigated Soils: As mentioned above, due to the ever growing pool of organic compounds discharged to the soil via wastewater, there is a general lack of knowledge regarding the effects that such substances cause to exposed organisms. In general terms, municipal wastewater is the main vector of OPECs to reach the environment, so that these substances are ubiquitous at sites where wastewater streams occur. Pharmaceutically active compounds (PACs) are designed to cause a defined effect on target organisms; however, when trace amounts of these substances are transported by wastewater into environment, they can interact with non–target organisms. One effect that has captured the attention of the scientific community in recent years is the development of antibiotic resistance by pathogenic microorganisms due to the occurrence of antibiotics in wastewater, surface water bodies and soils receiving wastewater [139–140]. However, a large number of studies on this subject report that proliferation of antibiotic–resistant pathogens is quite unlikely in wastewater irrigated soils [97–99]. Conversely, the study referenced in [96] attempts to relate the occurrence of sulfonamide and fluoroquinolone antibiotics with the emergence of antibiotic resistances in wastewater and long–term wastewater irrigated soils. The authors reported a relationship between time under irrigation and the frequency of detection of antibiotic resistance genes in soils. In the case of non–antibiotic PACs, the most studied compounds –because they are the most used worldwide– are the analgesic and anti–inflammatory drugs [141]. Compounds such as ibuprofen, naproxen, diclofenac, paracetamol and ketoprofen have been shown to cause systemic damages in aquatic species; damages in liver, gills and kidney are commonly reported [142]. The non–steroidal anti–inflammatory drug diclofenac has been demonstrated to cause visceral gout in vultures; in fact, the presence of diclofenac in livestock was the cause of the mass death of three species of vulture in India and Africa [143]. Other studies show that chronic exposure to traces of anti–inflammatory drugs leads to a lessening in the development of human embryo cells [144]. The occurrence of psychotropic agents at trace levels in water bodies polluted by wastewater discharges has been shown to alter the behavior of some fish species, suppressing their survival instincts against predators [145]. With regard to OPECs that are not pharmaceutically active compounds, there is significant concern that they may alter hormone homeostasis in organisms. These substances, known as endocrine disruptors, can mimic or compete with natural hormones by binding with active sites on hormone receptors, causing reduced or disproportionate hormonal responses in the affected organisms [146]. The most potent endocrine disruptors found so far in municipal wastewater are the natural and artificial estrogenic hormones –the latter are used as birth control agents– and the regulators of thyroidal function, followed by plasticizers (e.g. phthalates and bisphenols), surfactants and their metabolites and some industrial additives [147]. Endocrine disruptors are suspected of causing the feminization or masculinization of fish and reptile populations as well as the occurrence of breast cancer, imbalances in thyroidal function, teratogenic effects (e.g. cryptorchid) in mammals, and even obesity in mammals (obesogens) [148–150]. There is a serious lack of knowledge regarding the effects caused by OPECs in soil organisms. Studies on this field have been little developed compared to those for water bodies. Table 9 shows some examples of effects caused to soil organism by the occurrence of OPECs.

Table 9: Summary of negative effects on soil organisms caused by the occurrence of pollutants of emerging concern at trace levels

Compound	Effect in soil organisms	Reference
Estrone, 17 estradiol (hormones)	Negative impacts on the vegetative cycle of alfalfa (Medicago sativa).	[151]
Sex hormones	Shift in sex ratio of free life nematode communities in soil.	[152]
Triclosan (antibacterial agent)	Inhibition in plant growth (rice and cucumber). Effect concentrations 50 (EC50, i.e. 50% of exposed population was affected) were 57 and 108 mg/kg for rice and cucumber respectively. Inhibition of soil respiration and phosphatase activity at concentration levels higher than 10 mg/kg.	[153]
	Reduction in soil respiration 4 days after supplying the compound. The observed effects were dependent on the adsorption of the compoundZ onto the soil.	[154]
Bisphenol A (plasticizer)	Shift in sex ratio to female individuals in isopod (soil arthropod) communities.	[155]
Abamectin (anthelminthic)	Negative impacts on reproduction of Folsomia fimetaria and Folsomia candida (soil arthropods) at concentrations of 0.25 and 0.5 mg/kg of soil (dry mass), respectively. Negative impacts on reproduction of soil earthworms at concentration levels of 0.06 mg/kg of soil (dry mass).	[156]
Fenbendazole and cypermethrin (antiparasitic)	Negative impacts on degrading microorganisms of dung.	[157]

Sulfonamide and tetracycline antibiotics	Inhibition of the soil microbial activity by 10% (ED10) at concentrations of 0.003–7.35 µg/g of soil (dry mass). Shifts in fungi:bacteria ratio.	[158]
Sulfadiazine (antibiotic)	Decrease in denitrification rates when the input of antibiotic was 100 mg/kg of soil (dry mass).	[159]
	Significant decrease in the bacteria:fungi ratio	[160]
Chlortetracycline, tetracycline, tylosin, sulfamethoxazole, sulfamethazine and trimethoprim (antibiotics)	Decrease in crop growth (sweet oat, rice and cucumber). Inhibition of the microbial activity of soil (soil respiration and phosphatase enzyme activity).	[161]
Human and veterinary pharmaceutically active substances	Decrease in growth and development of Phaseolus vulgaris L., Glycine max, Medicago sativa, Zea mays, and several other crops.	[162]

The effects caused by this class of pollutants are not limited to soil organisms and impacts can be observed in the soil matrix. For example, surfactants can, on the one hand, decrease the capillarity and penetrability of soil as well as increase the solid–liquid contact angle, the shape factor and the sorptivity of soil particles. On the other hand, the input of these substances can increase the desorption of previously sorbed organic molecules on the soil particles, which in turn increases the bioavailability and mobility of the desorbed compounds [163]. To evaluate the toxic effects caused by the occurrence of OPECs to soil organisms two approaches are commonly used, i.e. acute and chronic toxicity studies. For the former, high concentrations of target pollutants are supplied to studied organisms under controlled conditions for a short period; chronic toxicity tests, on the other hand, are based on prolonged exposure of organisms to low (i.e. environmentally representative) doses of the studied pollutants. So far, most toxicity studies dealing with OPECs have been carried out using the acute toxicity approach. Even though these studies do not fully represent the conditions observed in the field, they provide valuable information on the subject of impacts caused by this kind of contaminants to soil organisms. Studies evaluating chronic toxic effects of pollutants are more representative of field conditions, i.e. toxic substances enter to soil in small doses over long periods. In this regard, conducting long–term toxicity studies that evaluate the chronic effects caused by OPECs in soil organisms are a priority. Several toxicity studies report that the effects of organic pollutants on soil organisms (i.e. reduction in soil respiration, enzymatic activity and nitrification/denitrification rates) are observed in the early days of exposition; then, after a short period (4 to 10 days) soil recovers to its basal conditions [153–154, 159, 161]. The next step in toxicity studies for these emerging pollutants is to determine the dynamics of the toxic effects on soil organisms after tens or hundreds of growing cycles in which target contaminants are continuously supplied; i.e. under conditions similar to what occurs in long–term irrigated areas.

Occurrence of Domestic Wastewater–Related Organic Pollutants in Irrigated Soils: In spite of the fact that wastewater is the main vehicle allowing OPECs to reach soil, very few studies reporting the presence of these pollutants in wastewater irrigated soils have been carried out. This finds an explanation, on the one hand, in the inherent difficulty of extracting and isolating organic compounds at trace levels from the soil matrix and, on the other hand, in the fact that analyzing this type of pollutants is relatively expensive. Figure 5 shows the sites where monitoring studies aimed at determining the occurrence of OPECs in wastewater irrigated soils have been performed. In this figure the number of sites monitored is contrasted with the 20 countries with the highest use of untreated wastewater for agricultural irrigation. Most of the monitoring studies are concentrated in China, the country using the highest volume of untreated wastewater in agriculture [17], followed by the United States and Mexico –the latter is the second placed country in terms of reuse of untreated wastewater for irrigation–.

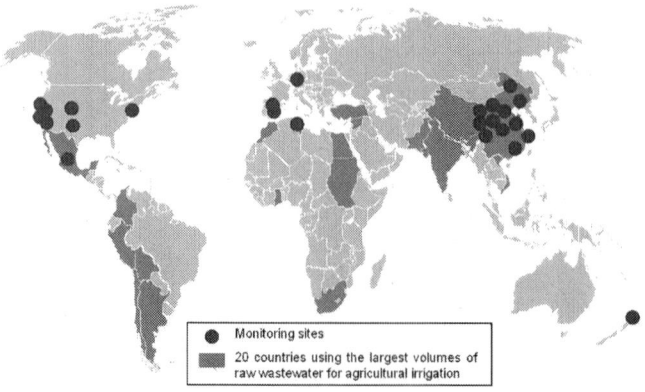

Figure 5: Monitoring studies for pollutants of emerging concern in wastewater irrigated soils throughout the world and comparison with the 20 countries using the largest volumes of raw wastewater for agricultural irrigation.

Efforts in monitoring emerging pollutants in developing countries where the use of raw wastewater is widespread are of value; this requires cooperation with research centers where analytical techniques are currently validated to perform soil analyses or by sharing "know how" and technology with developing countries in order to perform analysis on site. Determination of OPECs in soil requires an exhaustive extraction step, which in most cases has to be carried out at a moderately high temperature, particularly in the case of analysis of thermolabile compounds (e.g. sulfonamide antibiotics). Extraction methods such as pressurized fluids extraction, microwave assisted extraction and ultrasonic assisted extraction are preferred over traditional Soxhlet extraction techniques, since they guarantee greater contact between the solvent and the soil particles, resulting in higher recoveries of analytes. Analysis of OPECs is commonly accomplished using either liquid or gas chromatography techniques; although liquid chromatography is preferred as it is more suitable for the analysis of polar compounds, i.e. most PACs [164]. Monetary costs of these analyses are relatively high and analysis entails the use of potentially dangerous chemicals, which is in part the reason why monitoring studies for OPECs in soils are not carried out in poor countries. So far, the most reported emerging pollutants in wastewater irrigated soils are the pharmaceutically active substances (e.g. antibiotics, non–steroidal anti–inflammatory agents, anticonvulsants, anticoagulants and sex hormones), followed by plasticizers (e.g. phthalic acid esters and bisphenol A), metabolites of surfactants (e.g. nonylphenol, octylphenol) and antibacterial and antimycotic agents (e.g. triclosan and triclocarban). Table 10 shows the concentrations of some OPECs reported for wastewater irrigated areas.

Table 10. Concentrations of organic pollutants of emerging concern in treated/untreated wastewater irrigated soils

Compound	Concentration (µg/kg)	Comments	Reference
Carbamazepine	0.28–0.94	Concentration range observed in the surface layer (0–10 cm depth) of a treated wastewater irrigated soil during an irrigation cycle (May to October). The lowest concentration was observed before irrigation started while the highest concentration was determined in soil at the end of the irrigation cycle. Irrigation at the site has been occurring for the last 30 years.	[165–167]
	5.14 and 6.48	Concentrations found in the surface layer (0–10 cm depth) of Leptosol and Phaeozem soils, respectively that has been irrigated using untreated wastewater for 85 years.	
	4.92, 2.9 and 1.92	Concentrations found in forested, grass–covered and cultivated soil irrigated with treated wastewater for more than 25 years. Carbamazepine was found mainly in the first 30 cm of the soil profile.	

referenced in [182] found that non–steroidal anti–inflammatory drugs (NSAIDs) such as naproxen, which can produce negatively charged molecules after the ionization of the carboxyl functional group, are adsorbed to a lower extent than other compounds displaying higher hydrophobicity, such as carbamazepine or triclosan, in organic soils with high clay content. Organic compounds lacking of ionizable functional groups or displaying non–ionizable functional groups express their hydrophobicity by spontaneously migrating from water to the soil organic domain [183]. In wastewater irrigation systems, dissolved and particulate organic matter contained in wastewater tends to accumulate in the surface soil horizons, significantly favoring the build up of these compounds in topsoil. In studies referred to in [184] a greater accumulation of hydrophobic compounds, such as carbamazepine and esters of phthalic acid, was found in surface horizons of the irrigated soils, whereas hydrophilic compounds, namely ibuprofen, naproxen and diclofenac, were found in subsurface horizons. This behavior is explained, on the one hand, because hydrophilic compounds remain dissolved in water rather than being retained in soil and, on the other hand, because of hydrophilic compounds are more susceptible to desorption from soil either during further irrigation or heavy rain events, and thus tend to rapidly reach subsoil and the aquifer [180, 182].

Table 11: Relevant physical and chemical properties in terms of the environmental fate of emerging pollutants in soil

Polarity (ionization state at commonly found soil pH values)		
Positive	Negative	Positive/Negative (zwitterions)
Erythromycin (antibiotic)	Naproxen (non–steroidal analgesic drug)	Ofloxacin (antibiotic)
pKa 8.91	pKa 4.15	pKa 6.27 (COOH); pKa8.87 (NH2 +)
Hydrophobicity	Hydrophilic	
Hydrophobic	Ciprofloxacin hydrochloride (antibiotic)	
Di–2–(ethylhexyl)phthalate (plasticizer)		

Structure	Properties
	pKow 7.5 Solubility in water at 25°C: 4.1x10–2 g/L Volatility Volatile
Galaxolide (fragrance used in detergents) 	Vapor pressure: 7.27x10–2 Pa
	pKow – 0.82 Solubility in water at 25°C: 30 g/L Non-volatile
Bisphenol A (plasticizer precursor) 	Vapor pressure: 9.33x10–6 Pa

The chemical structure also affects the environmental fate of OPECs in soil. Molecules displaying aromatic moieties, such as carbamazepine and naproxen, have been shown to be strongly retained by soil organic matter –both to the aliphatic and aromatic fractions of soil organic matter– compared with compounds that have no resonance structures [182, 185–186]. This behavior is explained by the formation of bonds between aromatic rings within the solute molecules and the soil organic matter [185]. Nonylphenols and octylphenol compounds have surfactant properties as they possess an aliphatic chain and a phenol moiety at the edge of the molecule [177]. Due to this structure, these compounds can promote resolubilization of organic contaminants retained in soil, although the estimated risk of this occurrence is considerably low [187]. The presence of heteroatoms in organic molecules can impact upon their environmental fate in soil. For example, oxygen atoms within the ciprofloxacin molecule can form covalent bonds with aluminum and iron oxides in soil, resulting in irreversible adsorption of the compound onto the solid matrix [188]. With respect to volatile OPECs, artificial fragrances represent the best example of this feature; these compounds are contained both in personal care products and detergents. Typically, the more volatile compounds are also hydrophobic, so they can be spontaneously retained in topsoil and then volatilize when temperature increases [189]. Since irrigation using untreated wastewater, which contains large amounts of fragrances, is carried out in arid areas, it is expected that a significant fraction or all of the fragrance molecules are rapidly volatilized upon their input to soil via wastewater. Volatilization of OPECs in wastewater irrigated soils is still an unexplored issue; studies aimed at determining the fraction of organic contaminants that can be volatilized in soil enriched with organic matter via wastewater irrigation are still needed.

Natural attenuation processes leading to the removal and dissipation of OPECs in soil are shown inFigure 6. Contaminants may either dissipate in soil by photodegradation, biodegradation or chemical degradation (hydrolysis, oxidation or reduction) mechanisms; they may be accumulated in soil by adsorption or removed from soil by volatilization. There is a significant lack of information in the literature with regard to the natural photodegradation (i.e. photolysis of compounds by sunlight) of emerging pollutants in soil. The information available on the photodegradation of pesticides in agricultural soils is useful in elucidating, to some extent, the potential for photodegradation of OPECs in soil. Studies on natural photodegradation of the pesticide quinalphos showed that photodegradation takes place only in the first 2–5 mm of soil (photic layer); this photolysis takes place in two stages, each one at a different depth [190]. In the uppermost soil layer (the first 2 mm) direct photodegradation of the organic compounds (i.e. the transformation of compounds due to the direct incidence of photons) occurs; in this same layer, the production of free radicals (e.g. hydroxyl radicals and excited dissolved organic matter) occurs due to the breakdown of the soil organic matter. In the second stage, the free radicals migrate to the lower photic layer through facilitated transport by soil moisture; subsequently, photolytic transformation occurs below by the action of free radicals generated in the upper layer of soil (indirect photodegradation) [191]. As a result of the aforementioned aspects, soil moisture content as well as organic matter are determinant factors in the photodegradation of organic contaminants retained in the soil surface [191]. The physical structure of the soil can also significantly impact upon the photodegradation of organic pollutants, as it defines the depth to which solar radiation can penetrate the soil. In the study referenced in [192], sunlight photolysis of 4–nonylphenol in biosolids amended soils was studied. Photolysis resulted in 40% conversion of the compound within 30 days, with photodegradation observed in the first 5 mm of the soil. Since natural photodegradation occurs only in the soil surface layers, the organic compounds retained in the topsoil will be the most exposed to direct sunlight, although this does not necessarily imply increased rates of photodegradation. An example of this is the anticonvulsant agent carbamazepine, which is prominently retained in topsoil but has demonstrated poor photodegradation in water studies. Conversely, the anti–inflammatory drug diclofenac has been shown to present significant photoactivity [193], but it is less well retained in the topsoil. Due to this, photodegradation is unlikely to occur in soil by direct or indirect means. In spite of almost all of the studies evaluating the natural photodegradation of OPECs have been carried out in aqueous matrices [194], the results obtained in these experiments provide valuable information concerning the photoactivity of such compounds; which can be useful for studying the photolysis of organic pollutants in soil. For example, it is known that the NSAID ibuprofen and the anticonvulsants drugs carbamazepine and primidone are poorly photodegraded in water whereas the antibacterial agent triclosan, the antibiotic drug sulfamethoxazole and the NSAIDs diclofenac and naproxen are readily photodegraded [194]. These results can be the basis to establish experiments aimed at determining or modeling the photodegradation of organics in soil. In general terms, the natural sunlight reaching the troposphere (i.e. the surface of earth) does not possess enough energy to mineralize most photodegradable compounds [195]; therefore a wide variety of by–products occurs when organic contaminants in water and soil are photodegraded. It is known that, in some cases, more harmful compounds can be produced by photodegradation of some organic pollutants. For example, 2,8–dichlorodibenzo–p–dioxin is produced by the natural photodegradation of the antibacterial agent triclosan [196]. Differently to triclosan, its breakdown product has the potential to cause cancer in mammals. Another example is the antiepileptic drug carbamazepine, which photodegrades to acridone [197], a compound related to the occurrence of cancer in aquatic species.

Most photodegradation studies of OPECs (in water matrices) have been carried out in developed countries at latitudes higher than 30°N [194]. It is therefore necessary to investigate the intensity of photodegradation processes occurring at lower latitudes, in zones where higher incidence of sunlight occurs and treated/ untreated wastewater irrigation is more intensively practiced.

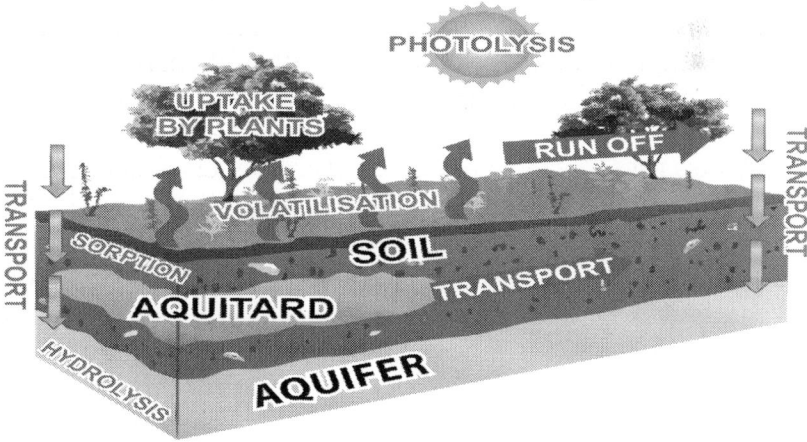

Figure 6: Processes involved in the environmental fate of emerging pollutants in soil.

Biodegradation of OPECs in soil has been studied in greater detail than photodegradation. Laboratory studies have found that biodegradation of emerging pollutants occurs optimally under aerobic conditions, while negligible transformations have been observed under anaerobic conditions [198]. This implies that biodegradation of this kind of contaminants is more likely to occur in well–structured soils, where tillage activities are frequently carried out, which allows better gas exchange through the soil matrix. The opposite behavior may be observed in anoxic/anaerobic soils, for instance in paddy fields. The antiepileptic drug carbamazepine is reported as one of the most refractory organic pollutants in soil, which has led researchers to consider this antiepileptic agent as a marker for anthropogenic contamination of surface and groundwater bodies [199]. In the study referenced in [200], mineralization of carbamazepine in soil was found to be less than 2%, after 120 days of incubation under aerobic conditions, while the reported in [201] show half–life times of 472 days in aerobic biosolids amended soils. Other compounds listed as recalcitrant in soil are the X–ray contrast media iopaminol, iomeprol and iohexol, whose biodegradation kinetic rate constants range from 0.29 to 0.46 µM/day [202]. Pharmaceuticals, such as the antiepileptic drug primidone and the psychoactive diazepam have shown recalcitrance in water [203]; further studies are necessary in order to elucidate whether such behavior may also occur in soil. Substances designed to exert an effect on microorganisms have been shown to be rapidly biodegraded in soil. Examples of these are antibacterial agents such as triclosan, triclocarban and antibiotic substances [204–205]. Triclosan and triclocarban have been shown to be biodegraded in aerobic soils after 18 and 108 days, respectively [198], whereas the antibiotic compounds erythromycin, oleandomycin, tylosin, tiamulin and salinomycin displayed half–lives in aerobic soil of 20, 27, 8, 16 and 5 days, respectively [205]. Endocrine disrupting compounds, such as phthalate esters have been shown to efficiently biodegrade in agricultural soils, displaying half–lives of 7.8 to 8.3 days for di-butyl phthalate and 26–30 days for di–2–(ethylhexyl)phthalate [206]. Currently, few soil microorganisms have been identified as degraders of emerging pollutants. For example, the fungi *Trametes versicolor* has been demonstrated to degrade naproxen [207], while *Rhodococcus rhodochorus* bacteria [208] have been shown to degrade carbamazepine down to levels of 15% of its initial concentration in soil. In the case of phthalates (plasticizers) bacteria belonging to groups of *Corynebacterium, Mycobacterium* and *Nocardia* were demonstrated to degrade up to 90% of di–butyl phthalate within 48 hours in biodegradation experiments using isolated bacteria cultivated in saline solution [209]. Knowing the species of microorganisms that perform the biodegradation of OPECs in soil is useful in order to design engineered systems to treat wastewater and polluted soils based on the increased ability of degraders to degrade specific compounds by acclimatization and bioaugmentation. Such systems were tested in [210] using the fungus *Trametes versicolor* to degrade up to 94% of carbamazepine in wastewater after 6 days in an air pulsed bed bioreactor. Biodegradation of OPECs in soil is influenced by the sorption phenomenon, therefore soil characteristics such as the content of organic matter, soil texture and soil pH are crucial for this process to occur. Adsorption of the organic contaminants onto the surface of the soil particles may favor biodegradation when the sorbed compounds are still bioavailable; conversely, when strong adsorption occurs (chemisorption on soil organic matter, clay or soil micropores) it can result in decreased bioavailability of the compounds and thus in the confinement of the pollutants within the soil matrix. Other properties of soil involved in the biodegradation of OPECs are: a) the climatic conditions of the site; b) the physical structure of soil; c) the soil moisture; and, d) the adaptation of soil organisms to biodegrade the target pollutants. It is possible that microorganisms in long–term wastewater irrigated soils more efficiently biodegrade OPECs than those living in non–irrigated soils or soils irrigated for a shorter time. This is due to the ability of soil microorganisms to adapt to using emerging pollutants as a carbon source. In this sense, studies comparing the degradation efficiency of OPECs in long–term wastewater irrigated soils with that observed in non–irrigated soils or newly irrigated soils are needed in order to establish appropriate strategies to prevent contamination of groundwater. Very few efforts have been made to determine the nature and quantity of by–products generated in soil by the biodegradation of OPECs. As shown in [200], biodegradation of emerging pollutants can generate by–products that can be more harmful than the original substance [196–197], thus the presence of these by–products as well as their environmental fate should be priority for further research.

Those emerging pollutants that are not degraded by soil microorganisms may either accumulate in soil, be assimilated by plants (if they are bioavailable) or be degraded by other mechanisms (e.g. photodegradation or hydrolysis). In the case of carbamazepine, studies referred to in [184] explain that this compound is one of the most highly accumulated in wastewater irrigated soils. Moreover, carbamazepine can be assimilated by plants in wastewater irrigation systems at environmentally relevant concentrations (i.e. within the range 1–3 µg/L). The study referenced in [211] shows that cucumber (*Cucumis sativus L.*) can accumulate carbamazepine in different parts of the plant: 4.5 µg/kg in roots, 1.9 µg/kg in stem, 39.9 µg/kg in leaves and 2.1 µg/kg in fruit. According to the authors, phytotoxic effects were observed when carbamazepine was supplied to soil by irrigation at concentrations as high as 10,000 µg/L. Results of this study show that consumption of carbamazepine polluted cucumber results in doses of 1 ng of carbamazepine per gram of fruit. Other studies show that soybean (*Glycine max L.*) can take up carbamazepine, triclosan and triclocarban in roots, stems and leaves at concentrations of 1.3–3.4 µg/kg for carbamazepine and 2.4–13.7 µg/kg for the antibacterial agent's triclosan and triclocarban. Concentrations of antibacterial agents in plants at a second harvesting were found to be higher than those obtained in the first one; this may be due to the accumulation of contaminants in the soil, as a bioavailable pool, between each irrigation events [212]. To date, the study of the assimilation of OPECs by plants in wastewater irrigated soils is still limited; moreover, priority should be given to develop health risk assessment studies related to the consumption of contaminated crops.

Adsorption (i.e. retention of solutes on the surface of the soil particles) of OPECs in the soil is a decisive process in their environmental fate, since through this process contaminants may either be retained or migrate into the aquifer. In cases where organic pollutants are retained in topsoil, photodegradation or volatilization phenomena can easily take place. The strength of the bonds that pollutants establish with soil particles determines the bioavailability of molecules to plants and soil microorganisms. Adsorption of pollutants onto soil is measured by the distribution coefficient (Kd) which relates the amount of compound retained by soil to the mass remaining in the liquid phase [213]. Several models to determine the distribution coefficient of organic compounds have been developed; such models vary in complexity and the accuracy with which they represent the field conditions; yet simple adsorption models such as linear, Langmuir and Freundlich are the most used [213]. Due to their organic nature, OPECs tend to be rapidly and strongly adsorbed by soil organic matter; due to this effect, non–polar emerging pollutants, such as phthalates, have been shown to instantly adsorb onto organic soils [214]. On the other hand, OPECs displaying negative charge at the soil pH values, as occurs for NSAIDs, exhibit less adsorption by soil due to the repulsive forces between the negatively charged moiety within the molecule and the soil particles displaying negative charges (i.e. organic matter and clay) [182]. Accumulation of organic matter in wastewater irrigated soils increases the soil's ability to adsorb organic compounds. The proof of this can be found in the study referenced in [96], which reports greater adsorption of the antibiotics sulfamethoxazole and ciprofloxacin in long–term wastewater irrigated soils compared to non–irrigated ones from the same area. In addition to soil organic matter, OPECs may be retained by the inorganic domain of soil; for instance, ciprofloxacin showed strong and instantaneous adsorption by iron oxides and clay in agricultural soils, which was achieved by the formation of covalent bonds between metals in the soil and the oxygen atoms within ciprofloxacin molecules [188]. Furthermore, adsorption of carbamazepine by smectite type clays has been reported by [215]. According to studies referred to in [182, 185, 215], the adsorption of OPECs with multiple aromatic rings is more efficient in soils displaying a high content of humified organic matter –which displays higher aromaticity than labile organic matter–. Polyaromatic compounds can establish π–π bonds between the aromatic rings within the pollutant molecules and aromatic compounds contained in soil organic matter. The formation of such bonds should be studied in future research in order to determine the optimum chemical characteristics of soil organic matter which enable better retention of contaminants, hence preventing their mobilization into the aquifer and/or making them available for uptake by plants. OPECs may be adsorbed by dissolved organic matter to soil via wastewater. Adsorption of organic pollutants to dissolved organic matter increases the solubility of the compounds and hence facilitates the lixiviation through soil. Studies referenced in [216–217] report that compounds such as naproxen, carbamazepine and sex hormones can be adsorbed onto dissolved organic matter, notably to the hydrophobic and neutral hydrophilic fractions of dissolved organic matter. The speed of formation and strength of bonds between organic compounds and the dissolved organic matter varies depending on the quality of both wastewater and dissolved organic matter in soil [217].

The continuous occurrence of OPECs in wastewater irrigated soils can impact upon the adsorption of other organic pollutants; this is because at the time emerging pollutants enter to soil via wastewater, some of the active adsorption sites in soil are still occupied by previously adsorbed pollutants. In the study referred to in [182], the distribution coefficients of three OPECs, namely naproxen, carbamazepine and triclosan, were determined by an adsorption model which takes into account the previous presence of organic pollutants in the soil (the initial mass model [218]). The authors found modest differences between the values obtained in their study and those reported in the literature. However, it was observed that compounds previously adsorbed onto soil, i.e. naproxen and carbamazepine, were released from the solid matrix each time wastewater "washes" the soil in each irrigation event, resulting in a risk of contamination of the aquifer.

The transportation of OPECs through soil is closely related to their adsorption onto the solid matrix. Transport studies can be performed using different approaches, either packed soil columns or undisturbed soil columns tests. Transport of OPECs and pathogens is better described using the undisturbed soil column approach; through this approach, it is possible to evaluate the impact of both physical and chemical properties of soil on the transport of pollutants. In transport assays using undisturbed soil columns it is possible to assess the impact of preferential paths on the transport of solutes and particles, at the same time determining the effect of chemical properties of soil in the retention of solutes under dynamic flow conditions. The type of clays in soil significantly impacts on the transport of organic pollutants. The presence of expansive clays in soil results in the disappearance of preferential paths in the porous network of soil once clay becomes wet, which in turn provokes the decay in transport of contaminants contained in water. However, in such cases, soil conditions become anaerobic and thus organic pollutants are biodegraded with difficulty. The understanding of the environmental fate of OPECs in wastewater irrigated soil still has many gaps. It is therefore important to carry out studies on the laboratory scale and then in the field (plot level or landscape level) in order to determine the fate of these substances under real conditions. Results of these studies are of great importance, on the one hand, to allow more accurate and useful risk assessment studies and, on the other hand, to determine the characteristics of the sites suitable for irrigation with treated/untreated wastewater without posing a risk to the health of organisms and to the quality of crops and water sources. Lastly, regulations for OPECs in soil should be established in order to set maximum concentration limits for the accumulation of these compounds in terms not only of the effects caused to soil organisms, but also their potential to reach groundwater.

PERSPECTIVES FOR FURTHER STUDIES

Reuse of wastewater in agricultural irrigation is a complex issue that requires the development of numerous studies in different disciplines; in this section some perspectives for further studies are presented.

- Long–term studies aimed at determining the improvement of soil properties to produce food. Such studies should compare the rate of entry and conversion of carbon, nitrogen and phosphorus in irrigated soils in order to obtain a mass balance showing either the sustainability or the accumulation of organic matter in wastewater irrigated soils. Moreover, studies demonstrating the long–term increase in the soil's ability to treat wastewater used for irrigation should be carried out for each of the properties addressed in this chapter, as well as those considered appropriate in each system.

- The determination of OPECs and pathogens in soils irrigated with wastewater. Such monitoring studies can be used to establish an inventory of contaminated sites that reflects the level of pollution in developed and developing countries. This can help in proposing *ad hoc* solutions for each site.

- Determining feasibility and the mechanisms that can lead to horizontal propagation of antibiotic resistance genes in soil microorganisms (either innocuous microorganisms or opportunistic pathogens).

- Chronic toxicity studies of OPECs in wastewater irrigated soils covering either several crop cycles, several generations of organisms or several years. Toxicity studies should address the effects of the presence of mixed contaminants at trace levels (environmentally relevant concentrations) on soil organisms. Such studies should be conducted including new emerging contaminants, e.g. nanoparticles.

- The study of the environmental fate of emerging contaminants using different model molecules in soil. Such environmental fate studies should be carried out at laboratory and field scale. In the case of environmental fate studies at laboratory scale, conditions used should be those that best emulate field conditions, e.g. sunlight lamp intensities similar to those observed in the field for testing photodegradation or undisturbed soil columns in transport assays through soil.

- The determination and quantification of the by-products appearing in soil upon dissipation of OPEC. Harmful compounds such as dioxins, chlorophenols and polyaromatics may be produced in soil from substances such as triclosan and carbamazepine. Discerning the occurrence and fate of these substances in soil should be addressed in future studies

- Determination of the environmental fate of organic, inorganic and microbial contaminants in agricultural soil remaining after irrigation with wastewater has ceased. Worldwide, notably in developed countries, there are several sites where irrigation with wastewater has been stopped after a considerable time; in such cases, it is necessary to know the fate of the pool of pollutants that accumulated in soil during continuous input via wastewater. Phenomena such as the release of heavy metals confined in soil organic matter can occur when soil organisms start to mineralize organic carbon accumulated in the soil. In addition, the soil microorganisms can lose the capacity to treat pollutants in wastewater, leaving the soil vulnerable in cases where wastewater irrigation is restarted.

- Studies elucidating the conditioning methods for agricultural soils newly irrigated with wastewater. Since in arid regions a considerable increase in the area under irrigation is being observed, it is necessary to use current knowledge to implement regulations establishing the optimal conditions for soils candidate to receive treated/untreated wastewater. These are necessary to prevent soil degradation and contamination of water sources in the irrigated area.

- Studies on the migration of contaminants in soil due to extreme events caused by climate change. Extreme rainfall events can cause an incremental increase in the mobilization of organic contaminants retained in the surface layers of soil into aquifers or to non–irrigated soils affected by runoff. However, increases in temperature can decrease the biodegradation of organic pollutants in the soil due to excessive drying of the solid matrix.

- The development and implementation of wastewater treatment systems to remove organic, inorganic and biological pollution without reducing the content of organic matter in the water. These systems must be inexpensive for dissemination in developing countries. Advanced primary treatment systems may represent a plausible strategy in such cases.

- The development and validation of environmentally–friendly analytical techniques for the determination of OPECs in soils.

CONCLUSIONS

The reuse of treated/untreated municipal wastewater for agricultural irrigation definitely has positive impacts on soil as a medium for the development of plants and animals; additionally, this practice results in positive impacts on the welfare of farmers due to the monetary savings and profits that they obtain by the use of wastewater as a fertilizer and water source for crops. Similarly, the soil's ability to self–cleanse and treat the wastewater supplied at each irrigation event increases with the reuse of wastewater. The accumulation of organic matter in the soil surface results in changes in soil pH to neutral and basic values, an improvement of soil physical structure and an increase in the soil microbial activity. Together with this, soil organisms become acclimatized to the presence of contaminants and thus their resilience to the harmful effects caused by pollutants increase. These phenomena lead to an improvement in the ability of the soil to act as a filter and transforming medium for contaminants and thereby to an increase in its capacity to treat wastewater. Such an improvement in soil functions can be capitalized by the State and the conventional treatment regime can be changed to a cheaper one driven by natural attenuation mechanisms. This in turn improves the quality of life of people living in the area by increasing food production and the possibility of obtaining profit by sales of produce. The responsible reuse of municipal wastewater for agricultural irrigation can help to mitigate three problems which are a priority in developing countries: a) water stress in arid areas where rain–

fed agriculture makes development uncertain. In such areas freshwater sources are used for agriculture rather than human consumption, and therefore the reuse of municipal wastewater not only results in savings of freshwater but also in the recharge of the aquifer in the irrigated area. Recharge is with good quality water produced by infiltration of wastewater through the soil; b) the food crisis and the lack of jobs in rural and peri–urban areas in developing countries. Reuse of wastewater represents a way of producing food for consumption and sale; and, c) the treatment of municipal wastewater generated in urban and rural areas through a low cost natural treatment systems which in turn generate profits for population.

In order to reuse wastewater responsibly and exploit its inherent benefits for soil and people living in the irrigated area, the occurrence of contaminants in wastewater –especially untreated wastewater– must be kept in mind. The presence of pathogenic microorganisms and the potential for antibiotic resistance dissipation via wastewater should be priority concerns in designing wastewater reuse schemes in agricultural areas, notably when using raw wastewater. Attention should be paid to the fate of emerging contaminants in wastewater irrigation schemes including its transportation through irrigation canals, storage in dams and deposition in agricultural soils and transport to aquifers. Another priority is the elucidation of the chronic toxic effects caused by the continuous presence of traces of emerging contaminants in irrigated soils. Since the group of OPECs is quite broad, model compounds should be selected to determine the rate at which they are dissipated or retained/transported through soil, as well the risk of these compounds reaching the aquifer or being assimilated by plants. Despite the spread of antibiotic resistance in the environment it has not been conclusively shown the role that irrigation with treated/untreated wastewater plays in this. To date, the concentrations of OPECs found in soil irrigated with wastewater are lower than the toxicity thresholds reported in literature. The precautionary principle states that wastewater must be minimally treated before irrigation in order to remove pathogenic microorganisms and trace of heavy metals, as well as to reduce as much as possible the concentration of emerging pollutants. Other areas of opportunity to be developed in order to reduce the risk of soil degradation and effects on soil organisms are: a) the development of environmentally friendly everyday–consumer products, containing organic compounds that have been proven to have no harmful effects on living organisms even at trace concentrations. Consumer products must follow strict risk assessments before release to the market; b) an improvement in health systems in cities in order to reduce the incidence of infectious diseases that ultimately generate biological contamination of soil, especially in irrigation systems using raw wastewater; c) the maintenance of wastewater irrigation schemes fed with municipal wastewater in order to avoid a high input of heavy metals and refractory organic compounds to soil and crops through irrigation; and, d) the *ad hoc* treatment of municipal wastewater to allow its reuse in agricultural activities. Low cost treatment systems aimed at removing microorganisms, suspended solids and trace heavy metals are recommended to treat wastewater without affecting its properties as a fertilizer and source of organic matter to improve physical, chemical and microbiological soil properties. Such an approach allows soil to fulfill its ecological functions as a generator of food and livelihoods and as a protective barrier to the aquifer.

REFERENCES

1. Elimelech M. The Global Challenge for Adequate and Safe Water. Aqua – Journal of Water Supply: Research and Technology 2006; 55(1) 3–10.

2. Rijsberman F.R. Water Scarcity: Fact or Fiction. Agricultural Water Management 2006;80 (1–3) 5–22.

3. United States Geological Survey. USGS: The USGS Water Science School. The World's Water – Distribution of Earth's Water. http://ga.water.usgs.gov/edu/earthwherewater.html (accessed on July 13 2013).

4. Shiklomanov I.A. World Water Resources – A New Appraisal and Assessment for the 21st Century. St Petersburg: UNESCO; 1998.

5. Food and Agriculture Organization of the United Nations, FAO. Coping with Water Scarcity: An Action Framework for Agriculture and Food Security. Rome: FAO; 2012.

6. Earth Trends: Environmental Information. http://www.wri.org/project/earthtrends/ (accessed on August 4th 2013).

7. Pereira L.S., Cordery I, and Iacovides I. Coping with Water Scarcity: Addressing the Challenges. Paris: UNESCO; 2009.

8.   DeFeo G., Mays L.W, and Angelakis A.N. Water and Wastewater Management Technologies in the Ancient Greece and Roman Civilizations. In: Wilder P. (ed.) Treatise in Water Science. Oxford: Elsevier; 2011. p3–22.

9.   Lazarova V., Levine B, and Sack J., Cirelli G., Jeffrey P., Muntau H., Salgot M., Brissaud F. Role of Water Reuse for Enhancing Integrated Water Management in Europe and Mediterranean Countries. Water Science and Technology 2001;43(10) 25–33.

10.   Anderson J. The Environmental Benefits of Water Recycling and Reuse. Water Science and Technology: Water Supply 2003;3(4) 1–10.

11.   Asano T. Wastewater Reclamation and Reuse: Water Quality Management Library. Pennsylvania: Technomic Publishing Company; 1998.

12.   Hussain I., Raschid–Sally L., Hanjra M.A., Marikar F, and van der Hoek W. Wastewater Use in Agriculture: Review of Impacts and Methodological Issues in Valuing Impacts. Colombo: International Water Management Institute, IWMI; 2002.

13.   Kretschmer N., Ribbe L, and Gaese H. Wastewater Reuse for Agriculture. Technology Resource Management and Development – Water Management 2002;2(1) 35–61.

14.   World Health Organization, WHO. Health Guidelines for the Use of Wastewater and Excreta in Agriculture. Geneve: WHO; 1989.

15.   Shelef G., Azov Y. The Coming Era of Intensive Wastewater Reuse in the Mediterranean Region. Water Science and Technology 1996;33(10–11) 115–125.

16.   Jiménez B. Wastewater Reuse to Increase Soil Productivity. Water Science and Technology 1995;32(12) 173–180.

17.   Jimenez B., Asano T. Water Reclamation and Reuse Around the World. In: Jimenez B., Asano T. (eds.) Water Reuse – An international Survey of Current Practice, Issues and Needs. London: IWA Publishing; 2008. p3–27.

18.   World Health Organization, WHO. Guidelines for the Safe Use of Wastewater, Excreta and Greywater. Geneva: WHO; 2006.

19.   Ensink J.H.J., Simmons R.W, and van der Hoek W. Wastewater Use in Pakistan: The Cases of Haroonabad and Faisalabad. In:

Scott C.A., Faruqui N.I., Raschid–Sally L. (eds.) Wastewater Use in Irrigated Agriculture: Confronting the Livelihood and Environmental Realities. Wallingford: CAB International; 2004. p91–99.

20. Toze S. Reuse of Effluent Water – Benefits and Risks. Agricultural Water Management 2006;80(1–3) 147–159.

21. Keraita B., Jiménez B, and Drechsel P. Extent and Implications of Agricultural Reuse of Untreated, Partially Treated and Diluted Wastewater in Developing Countries. CAB reviews: Perspectives in Agriculture, Veterinary Science, Nutrition and Natural Resources 2008;3(58) 1–15.

22. Cifuentes E., Gomez M., Blumenthal U., Tellez–Rojo M.M., Romieu I., Ruiz–Palacios G, and Ruiz–Velazco S. Risk Factors for Giardia Intestinalis Infection in Agricultural Villages Practicing Wastewater Irrigation in Mexico. American Journal of Tropical Medicine and Hygiene 2000;62(3) 388–392.

23. Mara D.D., Sleigh P.A., Blumenthal U.J, and Carr R.M. Health Risks in Wastewater Irrigation: Comparing Estimates from Quantitative Microbial Risk Analyses and Epidemiological Studies. Journal of Water and Health 2007;5(1) 39–50.

24. Baquero F., Martínez J.L, and Cantón R. Antibiotics and Antibiotics Resistance in Water Environments. Current Opinion in Biotechnology 2008;19(3) 260–265.

25. Knapp C.W., Dolfing J., Ehlert P.A.I, and Graham D.W. Evidence of Increasing Antibiotic Resistance Gene Abundances in Archived Soils Since 1940. Environmental Science and Technology 2010;44(2) 580–57.

26. Beausse J. Selected Drugs in Solid Matrices: A Review of Environmental Determination, Occurrence and Properties of Principal Substances. Trends in Analytical Chemistry 2004;23(10–11) 753–761.

27. Schmitt H and Römbke J. The Ecotoxicological Effects of Pharmaceuticals (Antibiotics and Antiparasiticides) in the Terrestrial Environment – A Review. In: Kúmmerer K. (ed.) Pharmaceuticals in the Environment: Sources, Fate, Effects and Risks. Heidelberg: Springer; 2008. p285–303.

28. Colborn T., vom Saal F.S, and Soto A.M. Developmental Effects of Endocrine–Disrupting Chemicals in Wildlife and Humans. Environmental Health Perspectives 1993;101(5) 378–384.

29. Daughton C.G. Non–Regulated Water Contaminants: Emerging Research. Environmental Impact Assessment Review 2004;24(7–8) 711–732.

30. Foppen J.W and Schijven J.F. Evaluation Data from the Literature on the Transport and Survival of Escherichia Coli and Thermotolerant Coliforms in Aquifers Under Saturated Conditions. Water Research 2006;40(3) 401–426.

31. Read D.S., Sheppard S.K., Bruford M.W., Glen D.M, and Symondson W.O.C. Molecular Detection and Predation by Soil–Arthropods on Nematodes. Molecular Ecology 2006;15(7) 1963–1972.

32. Berg G., Eberl L, and Hartmann A. The Rhizosphere as a Reservoir of Opportunistic Human Pathogen Bacteria. Environmental Microbiology 2005;7(11) 1673–1685.

33. Yamamoto H., Nakamura Y., Morigushi S., Nakamura Y., Honda Yuta., Tamura I., Hirata Y., Hayashi A, and Sekizawa J. Persistence and Partitioning of Eight Selected Pharmaceuticals in the Aquatic Environment: Laboratory Photolysis, Biodegradation and Sorption Experiments. Water Research 2009;43(2) 351–362.

34. Fausto Cereti C., Rossini F., Federici F., Quaratino D., Vassilev N, and Fenice M. Reuse of Microbially Treated Olive Mill Wastewater as Fertiliser for Wheat (Triticum Durum). Bioresource Technology 2004;91(2) 135–140.

35. Bradford A., Brook R, and Hunshal C. Wastewater irrigation: Hubli–Dharwad, India. International Symposium on Water, Poverty and Productive uses of Water at the Household Level, 21–23 January 2003, Muldersdrift, South Africa.

36. Van der Hoek W., Ul–Hassan M., Ensink J.H.J., Feenstra S., Raschid–Sally L., Munir S., Aslam R., Ali N., Hussain R, and Matsuno Y. Urban Wastewater: A Valuable Resource for Agriculture. Colombo: International Water Management Institute Research; 2002.

37. Gaye M., Niang S. Epuration des eaux usées et l'agriculture urbaine – Etudes et Recherches. Dakar: ENDA–TM; 2002.

38. Ramírez–Fuentes E., Lucho–Constantino C., Escamilla–Silva E, and Dendooven L. Characteristics and Carbon and Nitrogen Dynamics in Soil Irrigated with Wastewater for Different Lengths of Time. Bioresource Technology 2002;85(2) 179–187.

39. Fan A.M., Steinberg V.E. Health Implications of Nitrate and Nitrite in Drinking Water: an Update on Methemoglobinemia Occurrence and Reproductive and Developmental Toxicity. Regulatory toxicology and pharmacology 1996;23(1) 35–43.

40. Girovich M. Biosolids Treatment and Management: Processes for Beneficial Use. New York: Marcel Dekker Inc; 1996.

41. Degens B., Schipper L., Claydon J., Russell J, and Yeates G. Irrigation of an Allophanic Soil with Dairy Factory Effluent for 22 Years: Responses of Nutrient Storage and Soil Biota. Australian Journal of Soil Research. 2000;38(1) 25–35.

42. Mikkelsen R and Camberato J. Potassium, Sulfur, Lime and Micronutrient Fertilizers. In: Rechcigl J. E. (ed.) Soil Amendments and Environmental Quality. Boca Raton: CRC Press; 1995. p.109–137.

43. Sala L and Serra M. Towards Sustainability in Water Recycling. Water Science and Technology 2004;50(2) 1–8.

44. Steen I. Phosphorus Availability in the 21st Century: Management of a Non–Renewable Resource. Phosphorus and Potassium 1998;217(1) 25–31.

45. Jiménez B and Chávez A. Treatment of Mexico City Wastewater for Irrigation Purposes. Environmental Technology 1997;18(7) 721–729.

46. Coleman D.C., Crossley J.D.A., Hendrix P.F. Fundamentals of soil ecology. London: Academic Press; 2004.

47. Lal R. Soil Structure and Sustainability. Journal of Sustainable Agriculture 1991;1(4) 67–92.

48. Pardo A., Amato M, and Quaglietta Chiaranda F. Relationships between Soil Structure, Root Distribution and Water Uptake of Chickpea (Cicer arietinum L.). Plant Growth and Water Distribution. European Journal of Agronomy 2000;13(1) 39–45.

49. Boix–Fayos C., Calvo–Cases A., Imeson A.C., Soriano–Soto M.D, and Tiemessen I.R. Spatial and Short Term Variations in Runoff, Soil Aggregation and Other Soil Properties Along a Mediterranean Climatological Gradient. Catena 1998;33(2) 123–138.

50. Edwards A.P. Bremner J.M. Microaggregates in Soil. European Journal of Soil Science 1967;18(1) 64–73.

51. Bronick C.J and Lal R. Soil Structure and Management. Geoderma 2005;124(1–2) 3–22.

52. Martens D.A and Frankenberger W.T. Modification of Infiltration Rates in an Organic–Amended Irrigated. Agronomy Journal 1992;84(4) 707–717.

53. Armstrong A.S.B and Tanton T.W. Gypsum Application to Aggregated Saline–Sodic Clay Topsoils. Journal of Soil Science 1992;43(2) 249–260.

54. Jastrow J.D. Soil Aggregate Formation and the Accrual of Particulate and Mineral–Associated Organic Matter. Soil Biology and Biochemistry 1996;28(4–5) 665–676.

55. Haynes R.J and Naidu R. Influence of Lime, Fertilizer and Manure Applications on Soil Organic Matter Content and Soil Physical Structure: A Review. Nutrient Cycling in Agroecosystems 1998;51(2) 123–137.

56. Czarnes S., Hallett, P.D., Bengough A.G, and Young I.M. Root– and Microbial–Derived Mucilages Affect Soil Structure and Water Transport. European Journal of Soil Science 2000;51(3) 435–443.

57. Haynes R.J anb Beare M.H. Influence of Six Crops Species on Aggregate Stability and Some Labile Organic Matter Fractions. Soil Biology and Biochemistry 1997;29(11–12) 1647–1653.

58. Raimbault B.A and Vyn T.J. Crop Rotation and Tillage Effects on Corn Growth and Soil Structural Stability. Agronomy Journal 1991;83(6) 979–985.

59. Brunetti G., Senesi N, and Plaza C. Effects on Amendments with Treated and Untreated Olive Oil Mill Wastewaters on Soil Properties, Soil Humic Substances and Wheat Yield. Geoderma 2007;138(1–2) 144–152.

60. Martens D.A. Plant Residue Biochemistry Regulates Soil Carbon Cycling and Carbon Sequestration. Soil Biology and Biochemistry 2000;32(3) 361–369.

61. Cox L., Celis R., Hermosin M.C., Becker A, and Cornejo J. Porosity and Herbicide Leaching in Soils Amended with Olive–Mill Wastewater. Agriculture, Ecosystem and Environment 1997;65(2) 151–161.

62. Coppola A., Santini A., Botti P., Vacca S., Comegna V, and., Severino G. Methodological Approach for Evaluating the Response of Soil Hydrological Behavior to Irrigation with Treated Municipal Wastewater. Journal of Hydrology 2004;292(1–4) 114–134.

63. Wang Z., Chang A.C., Wu L, and Crowley D. Assessing the Soil Quality of Long–Term Reclaimed Wastewater–Irrigated Cropland. Geoderma 2003;114(3–4) 261–278.

64. Tam N.F.Y. Effects of Wastewater Discharge on Microbial Populations and Enzyme Activities in Mangrove Soils. Environmental Pollution 1998;102(2–3) 233.242.

65. Friedel J.K., Langer T., Siebe C, and Stahr K. Effects of Long–Term Waste Water Irrigation on Soil Organic Matter, Soil Microbial Biomass and Its Activities in Central Mexico. Biology and Fertility of Soils 2000;31(5) 414–421.

66. Filip Z., Kanazawa S., Berthelin J. Characterization of Effects of a Long–Term Wastewater Irrigation on Soil Quality by Microbiological and Biochemical Parameters. Journal of Plant Nutrition and Soil Science 1999;162(4) 409–413.

67. Kay B.D. Soil Structure and Organic Carbon: A Review. In: Lal R., Kimble J.M., Follett R.F., Stewart B.A. (eds.) Soil Processes and the Carbon Cycle. Boca Raton: CRC Press; 1998. p169–197.

68. Androver M., Farrus E., Moya G., Vadell J. Chemical Properties and Biological Activity in Soils of Mallorca Following Twenty Years of Treated Wastewater Irrigation. Journal of Environmental Management 2012;95 188–192.

69. Ansari M.I., Malik A. Biosoption of Nickel and Cadmium by Metal Resistant Bacterial Isolates from Agricultural Soil Irrigated with Industrial Wastewater. Bioresource Technology 2007;98(16) 3149–3153.

70. Kim D.Y., Burger J.A. Nitrogen Transformations and Soil Processes in a Wastewater–Irrigated, Mature Appalachian Hardwood Forest. Forest Ecology and Management 1997;90(1) 1–11.

71. Russell J.M., Cooper R.N., Lindsey S.B. Soil Denitrification Rates at Wastewater Irrigation Sites Receiving Primary–Treated and Anaerobically Treated Meat–Processing Effluent. Bioresource Technology 1993;43(1) 41–46.

72. Bouwer H. Groundwater Recharge with Sewage Effluent. Water Science and Technology 1991; 23(10–12) 2099–2108.

73. Bouwer H. Soil–Aquifer Treatment of Sewage. Rome: Food and Agriculture Organization of the United Nations; 1987.

74. Quanrud D.M., Hafer J., Karpiscak M.M., Zhang J., Lansey K.E., Arnold R.G. Fate of Organics During Soil–Aquifer Treatment: Sustainability of Removals in the Field. Water Research 2003;37(14) 3401–3411.

75. Pescod M. Wastewater Treatment and Use in Agriculture, FAO irrigation and drainage Paper 47. Rome: Food and Agriculture Organization; 1992.

76. Ayres D., Westtcott W. Water Quality for Agriculture, FAO Paper 29. Rome: Food and Agricultural Organization; 1985.

77. Arora M., Kiran B., Rani S., Rani A., Kaur B., Mittal N. Heavy Metal Accumulation in Vegetables Irrigated with Water from Different Sources. Food Chemistry 2008;111(4) 811–815.

78. Idelovitch E., Michail M. Soil–Aquifer Treatment: A New Approach to an Old Method for Wastewater Reuse. Journal (Water Pollution Control Federation) 1984;56(8) 936–943.

79. Lin C., Shacahr Y., Banin A. Lin, C. Heavy Metal Retention and Partitioning in a Large–Scale Soil–Aquifer Treatment (SAT) System Used for Wastewater Reclamation. Chemosphere 2004;57(9) 1047–1058.

80. Gadd G.M. Microbial Influence on Metal Mobility and Application for Bioremediation. Geoderma 2004;122(2–4) 109–119.

81. Kunhikrishnan A., Bolan N.S., Müller K., Laurenson S., Naidu R., Kim W.I. The Influence of Wastewater Irrigation on the Transformation and Bioavailability of Heavy Metal(loid)s in Soil. In: Sparks D.L. (ed.) Advances in Agronomy, Vol. 115. London: Academic Press; 2013. p.219–273.

82. Aleem A., Isar J., Malik A. Impact of Long–Term Application of Industrial Wastewater on the Emergence of Resistance Traits in Azotobacter chroococcum Isolated from Rhizospheric Soil. Bioresource Technology 2003;86(1) 7–13.

83. Altaf M.M., Masood F., Malik A. Impact of Long–Term Application of Treated Tannery Effluents on the Emergence of Resistance Traits in Rhizobium sp. Isolated from Trifolium Alexandrinum. Turkish Journal of Biology 2008;32(1) 1–8.

84. Müller K., Duwig C., Prado B., Siebe C., Hidalgo C., Etchevers J. Impact of Long–Term Wastewater Irrigation on Sorption and Transport of Atrazine in Mexican Agricultural Soils. Journal of Environmental Science and Health, Part B 2012;47(1) 30–41.

85. von Oepen B., Kördel W., Klein W. Sorption of Nonpolar and Polar Compounds to Soils: Processes, Measurements and Experience with the Applicability of the Modified OECD–Guideline 106. Chemosphere 1991;22(3) 285–304.

86. Müller K., Magesan G.N., Bolan N.S. A Critical Review of the Influence of Effluent Irrigation on the Fate of Pesticides in Soil. Agriculture, Ecosystems and Environment 2007;120(2) 93–116.

87. Landa–Cansigno O., Durán–Álvarez J.C., Jiménez B. Retention of Escherichia coli, Giardia lamblia Cysts and Ascaris lumbricoides Eggs in Agricultural Soils Irrigated by Untreated Wastewater. Journal of Environmental Management 2013;128 22–29.

88. Shuval H., Adin A., Fattal B., Rawutz E., Yekutiel P. Wastewater Irrigation in Developing Countries: Health Effects and Technical Solutions – Technical Paper No. 51. Washington: The World Bank; 1986.

89. Seymour I.J., Appleton H. Foodborne Viruses and Fresh Produce. Journal of Applied Microbiology 2001;91(5) 759–773.

90. [90] Bos R., Carr R., Keraita B. Assessing and Mitigating Wastewater–Related Health Risks in Low–Income Countries: An Introduction. In: Drechsel Pay, Scott C.A., Raschid–Sally L., Redwood M., Bahri A. (eds) Wastewater Irrigation and Health: Assessing and Mitigating Risk in Low–Income Countries. London/Ottawa/Colombo: Earthscan/IDRC/IWMI; 2010. p.29–51.

91. Jiménez B., Drechsel P., Koné D., Bahri A., Raschid–Sally L., Qadir M. General Wastewater, Sludge and Excreta Use Situation. In: Drechsel Pay, Scott C.A., Raschid–Sally L., Redwood M., Bahri A. (eds) Wastewater Irrigation and Health: Assessing and Mitigating Risk in Low–Income Countries. London/Ottawa/Colombo: Earthscan/IDRC/IWMI; 2010. p.3–29.

92. Barwick R.S., Mohammed H.O., White M.E., Bryant R.B. Prevalence of Giardia spp. and Cryptosporidium spp. on Dairy Farms in Southeastern New York State. Preventive Veterinary Medicine 2003;59(1–2) 1–11.

93. Gupta N., Khan D.K., Santra S.C. Prevalence of Intestinal Helminth Eggs on Vegetables Grown in Wastewater–Irrigated Areas of Titagarh, West Bengal, India. Food control 2009;20(10) 942–945.

94. [94] Rizzo L., Manaia C., Merlin C., Schwartz T., Dagot C., Ploy M.C., Michael I., Fatta–Kassinos D. Urban Wastewater Treatment Plants as Hotspots for Antibiotic Resistant Bacteria and Genes Spread into the Environment: A Review. Science of the Total Environment 2013;447(1) 345–360.

95. Berg G., Marten P., Ballin G. Stenotrophomonas maltophilia in the Rhizosphere of Oilseed Rape – Occurrence, Characterization and Interaction with Phytopathogenic Fungi. Microbiology Research 1996;151(1) 19–27.

96. Dalkmann P., Broszat M., Siebe C., Willaschek E., Sakinc T., Huebner J., Amelung W., Grohmann E., Siemens J. Accumulation of Pharmaceuticals, Enterococcus, and Resistance Genes in Soils Irrigated with Wastewater for Zero to 100 Years in Central Mexico. PLoS ONE 2012;7(9) e45397.

97. Negreanu Y., Pasternak Z., Jurkevitch E., Cytryn E. Impact of Treated Wastewater Irrigation on Antibiotic Resistance in Agricultural Soils. Environmental Science and Technology 2012;46(9) 4800–4808.

98. McLain J.E.T., Williams C.F. Development of antibiotic resistance in bacteria of soils irrigated with reclaimed wastewater. In: proceedings of the 5th National Decennial Irrigation Conference, 5–8 December 2010, Phoenix, USA. 2010.

99. Gatica J., Cytryn E. Impact of Treated Wastewater Irrigation on Antibiotic Resistance in the Soil Microbiome. Environmental Science and Pollution Research 2013;20(6) 3529–3538.

100. Jiménez B. Wastewater Use in Agriculture: Public Health Considerations. In: Trimble T.W., Trimble S.W. (eds.) Encyclopedia of Water Science. London: CRC Press; 2007. p.1303–1306.

101. Oron G., DeMalach Y., Hoffman Z., Manor Y. Effect of Effluent Quality and Application Method on Agricultural Productivity and Environmental Control. Water Science and Technology 1992;26(7–8) 1593–1601.

102. Najafi P., Mousavi S., Feizi M. Effects of Using Treated Municipal Wastewater in Irrigation of Tomato. Journal of Agricultural Science and Technology 2003;15 (1) 65–72.

103. Patel J., Millner P., Nou X., Sharma M. Persistence of Enterohaemorrhagic and Nonpathogenic E. coli on Spinach Leaves and in Rhizosphere Soil. Journal of Applied Microbiology 2010;108(5) 1789–1796.

104. Heaton J.C., Jones K. Microbial Contamination of Fruit and Vegetables and the Behaviour of Enteropathogens in the Phyllosphere: A Review. Journal of Applied Microbiology 2008;104(3) 613–626.

105. Feachem R.G., Bradley D.J., Garelick H., Mara D.D. Sanitation and Disease – Health Aspects of Excreta and Wastewater Management. Chichester: The World Bank; 1983.

106. Strauss M. Human Waste (Excreta and Wastewater). Swiss Federal Institute of Aquatic Science and Technology (EAWAG). http://www.eawag.ch/organisation/abteilungen/sandec/publikationen/publications_wra/downloads_wra/human_waste_use_ETC_SIDA_UA.pdf (accessed on August 20 2013).

107. Solomon E.B., Yaron S., Matthews K.R. Transmission of Escherichia coli O157:H7 from Contaminated Manure and Irrigation Water to Lettuce Plant Tissue. Applied and Environmental Microbiology 2002;68(1) 397–400.

108. Zhang G., Ma L., Beuchat L.R., Erickson M.C., Phelan V.H., Doyle, M.P. Lack of Internalization of Escherichia coli O157:H7 in Lettuce (Lactuca sativa L.) After Leaf Surface and Soil Inoculation. Journal of Food Protection 2009;72(10) 2028–2037.

109. Mitra R., Cuesta–Alonso E., Wayadande A., Talley J., Gilliland S., Fletcher J. Effect of Route of Introduction and Host Cultivar on the Colonization, Internalization, and Movement of the Human Pathogen Escherichia coli O157:H7 in Spinach. Journal of Food Protection 2009;72(7) 1521–1530.

110. Guo X., Chen J., Brackett R.E., Beuchat L.R. Survival of Salmonellae on and in Tomato Plants from the Time of Inoculation at Flowering and Early Stages of Fruit Development Through Fruit Ripening. Applied and Environmental Microbiology 2001;67(10) 4760–4764.

111. Barker–Reid F., Harapas D., Engleitner S., Kreidl S., Holmes R., Faggian R. Persistence of Escherichia coli on Injured Iceberg Lettuce in the Field, Overhead Irrigated with Contaminated Water. Journal of Food Protection 2009;72(3) 458–464.

112. Amahmid O., Asmama S., Bouhoum K. The Effect of Waste Water Reuse in Irrigation on the Contamination Level of Food Crops by Giardia Cysts and Ascaris Eggs. International Journal of Food Microbiology 1999;49(1) 19–26.

113. Erdogrul Ö., Şener H. The Contamination of Various Fruit and Vegetable with Enterobius vermicularis, Ascaris Eggs, Entamoeba histolyca Cysts and Giardia Cysts. Food Control 2005;16(6) 557–560.

114. Ensink J.H., Mahmood T., Dalsgaard A. Wastewater-Irrigated Vegetables: Market Handling Versus Irrigation Water Quality. Tropical Medicine and International Health 2007;12(s2) 2–7.

115. Gomes C., Da Silva P., Moreira R.G., Castell–Perez E., Ellis E., Pendleton M. Understanding E. coli Internalization in Lettuce Leaves for Optimization of Irradiation Treatment. International Journal of Food Microbiology 2009;135(3) 238–247.

116. Chen M. Pollution of Ground Water by Nutrients and Fecal Coliforms from Lakeshore Septic Tank Systems. Water, Air, and Soil Pollution 1988;37(3–4) 407–417.

117. Stewart L.W., Reneau R.B. Spatial and Temporal Variation of Fecal Coliform Movement Surrounding Septic Tank–Soil Absorption Systems in Two Atlantic Coastal Plain Soils. Journal of Environmetal Quality 1981;10(4) 528–531.

118. Keswick B.H., Gerba, C.P. Viruses in Groundwater. Environmental Science and Technology 1980;14(11) 1290–1297.

119. Gerba C.P., Melnick J.L., Wallis C. Fate of Wastewater Bacteria and Viruses in Soil. Journal of the irrigation and drainage division 1975;101(3) 157–174.

120. Smith M.S., Thomas G.W., White R.E., Ritonga D. Transport of Escherichia coli Through Intact and Disturbed Soil Columns. Journal of Environmental Quality 1985;14(1) 87–91.

121. Abu–Ashour J., Joy D.M., Lee H., Whiteley H.R., Zelin S. Transport of Microorganisms Through Soil. Water, Air, and Soil Pollution 1994;75(1–2) 141–158.

122. Gerba C.P., Bitton G. Microbial Pollutants: Their Survival and Transport Pattern to Groundwater. In: Bitton G., Gerba C.P. (eds.) Groundwater Pollution Microbiology. New York: John Wiley & Sons, Inc.; 1984. p.65–88.

123. Gannon J.T., Manilal V.B., Alexander M. Relationship Between Cell Surface Properties and Transport of Bacteria Through Soil. Applied and environmental microbiology 1991;57(1) 190–193.

124. Scholl M.A., Mills A.L., Herman, J.S., Hornberger, G.M. The Influence of Mineralogy and Solution Chemistry on the Attachment of Bacteria to Representative Aquifer Materials. Journal of Contaminant Hydrology 1990;6(4) 321–336.

125. Hsu B.M., Huang C. Influence of Ionic Strength and pH on Hydrophobicity and Zeta Potential of Giardia and Cryptosporidium. Colloids and Surfaces A: Physicochemical and Engineering Aspects 2002;201(1) 201–206.

126. Gannon J., Tan Y.H., Baveye P., Alexander, M. Effect of Sodium Chloride on Transport of Bacteria in a Saturated Aquifer Material. Applied and Environmental Microbiology 1991;57(9) 2497–2501.

127. Gaspard P.G., Wiart J., Schwartzbrod J. Étude Expérimentale de l'Adhésion des Ceufs d'Helminthes Ascaris suum: Consequences pour l'Environnent. Revue des Sciences de L'Eau 7, 367–376.

128. Chang A.C., Page A.L., Asano T. Developing Human Health–Related Chemical Guidelines for Reclaimed Wastewater and Sewage Sludge Applications in Agriculture. Geneva: World Health Organization; 1995.

129. World Health Organization, WHO Environmental Health Criteria 135: Cadmium–Environmental Aspects. Geneva: World Health Organization; 1992.

130. Yuan Y. Etiological Study of High Stomach Cancer Incidence Among Residents in Wastewater Irrigated Areas. Environmental Protection Science 1993;19(1) 70–73.

131. Nicholson F.A., Smith S.R., Alloway B.J., Carlton–Smith C., Chambers B.J An Inventory of Heavy Metals Inputs to Agricultural Soils in England and Wales. Science of the Total Environment 2003;311(1) 205–219.

132. Siebe C., Cifuentes E. Environmental Impact of Wastewater Irrigation in Central Mexico: An Overview. International Journal of Environmental Health Research 1995;5(2) 161–173.

133. Leach L., Enfield C., Harlin C. Summary of Long–Term Rapid Infiltration System Studies. EPA Report EPA–600/2–80–165. Oklahoma: U.S. EPA; 1980.

134. Mapanda F., Mangwayana E.N., Nyamangara J., Giller K.E. The Effect of Long–Term Irrigation Using Wastewater on Heavy Metal Contents of Soils Under Vegetables in Harare, Zimbabwe. Agriculture, Ecosystems & Environment 2005;107(2) 151–165..

135. Muchuweti M., Birkett J.W., Chinyanga E., Zvauya R., Scrimshaw M.D., Lester J.N. Heavy Metal Content of Vegetables Irrigated with Mixtures of Wastewater and Sewage Sludge in Zimbabwe: implications for Human Health. Agriculture, Ecosystems & Environment 2006;112(1) 41–48.

136. Barceló D. Emerging Pollutants in Water Analysis. Trends in Analytical Chemistry 2003;22(10) xiv–xvi.

137. Kümmerer K. Pharmaceuticals in the Environment: Sources, Fate, Effects and Risks. Berlin: Springer–Verlag; 2009.

138. Bolong N., Ismail A.F., Salim M.R., Matsuura T. A Review of the Effects of Emerging Contaminants in Wastewater and Options for Their Removal. Desalination 2009;239(1) 229–246.

139. Kümmerer K. Significance of Antibiotics in the Environment. Journal of Antimicrobial Chemotherapy, 2003;52(1), 5–7.

140. Thiele-Bruhn S. Pharmaceutical Antibiotic Compounds in Soils – A Review. Journal of Plant Nutrition and Soil Science 2003;166(2) 145–167.

141. Santos L.H., Araújo A.N., Fachini A., Pena A., Delerue–Matos C., Montenegro M.C.B.S.M. Ecotoxicological Aspects Related to the Presence of Pharmaceuticals in the Aquatic Environment. Journal of Hazardous Materials 2010;175(1) 45–95.

142. Fent K., Weston A.A., Caminada D. Ecotoxicology of Human Pharmaceuticals. Aquatic Toxicology 2006;76(2) 122–159.

143. Swan G.E., Cuthbert R., Quevedo M., Green R.E., Pain D.J., Bartels P., Cunningham A.A., Duncan N., Meharg A.A., Oaks J.L., Parry–Jones J., Shultz S., Taggart M.A., Verdoorn G., Wolter K. Toxicity of Diclofenac to Gyps Vultures. Biology Letters 2006;2(2) 279–282.

144. Pomati F., Castiglioni S., Zuccato E., Fanelli R., Vigetti D., Rossetti C., Calamari D. Effects of a Complex Mixture of Therapeutic Drugs at Environmental Levels on Human Embryonic Cells. Environmental Science and Technology 2006;40(7) 2442–2447.

145. Mennigen J.A., Stroud P., Zamora J.M., Moon T.W., Trudeau V.L. Pharmaceuticals as Neuroendocrine Disruptors: Lessons Learned from Fish on Prozac. Journal of Toxicology and Environmental Health, Part B 2011;14(5–7) 387–412.

146. Welshons W.V., Thayer K.A., Judy B.M., Taylor J.A., Curran E.M., Vom Saal F.S. (2003). Large Effects from Small Exposures. I. Mechanisms for Endocrine–Disrupting Chemicals with Estrogenic Activity. Environmental Health Perspectives 2003;111(8) 994–1006.

147. Petrovic M., Eljarrat E., De Alda M.L., Barceló D. Endocrine Disrupting Compounds and Other Emerging Contaminants in the Environment: A Survey on New Monitoring Strategies and Occurrence Data. Analytical and Bioanalytical Chemistry 2004;378(3) 549–562.

148. Bigsby R., Chapin R.E., Daston G.P., Davis B.J., Gorski J., Gray L.E., Howdeshell K.L., Zoeller R.T., vom Saal, F.S. Evaluating the Effects of Endocrine Disruptors on Endocrine Function During Development. Environmental Health Perspectives 1999;107(4) 613–618.

149. Grün F., Blumberg B. Endocrine Disrupters as Obesogens. Molecular and Cellular Endocrinology 2009;304(1) 19–29.

150. Birnbaum L.S., Fenton, S.E. Cancer and Developmental Exposure to Endocrine Disruptors. Environmental health perspectives 2003;111(4) 389–394.

151. Shore L.S., Kapulnik Y., Ben-Dor B., Fridman Y., Wininger S., Shemesh M. Effects of Eestrone and 17 β-Estradiol on Vegetative Growth of Medicago sativa. Physiologia Plantarum 1992;84(2) 217–222.

152. Hu C., Hermann G., Pen–Mouratov S., Shore L., Steinberger Y. Mammalian Steroid Hormones Can Reduce Abundance and Affect the Sex Ratio in a Soil Nematode Community. Agriculture, Ecosystems and Environment 2011;142(3) 275–279.

153. Liu F., Ying G.G., Yang L.H., Zhou Q.X. Terrestrial Ecotoxicological Effects of the Antimicrobial Agent Triclosan. Ecotoxicology and Environmental Safety 2009;72(1) 86–92.

154. Butler E., Whelan M.J., Sakrabani R., van Egmond R. Fate of Triclosan in Field Soils Receiving Sewage Sludge. Environmental Pollution 2012;167 101–109.

155. Lemos M.F., van Gestel C.A., Soares A.M. Endocrine Disruption in a Terrestrial Isopod Under Exposure to Bisphenol A and Vinclozolin. Journal of Soils and Sediments 2009;9(5) 492–500.

156. Diao X., Jensen J., Hansen A.D. Toxicity of the Anthelmintic Abamectin to Four Species of Soil Invertebrates. Environmental Pollution 2007;148(2) 514–519.

157. Sommer C., Bibby B.M. The Influence of Veterinary Medicines on the Decomposition of Dung Organic Matter in Soil. European Journal of Soil Biology 2002;38(2) 155–159.

158. Thiele–Bruhn S., Beck I.C. Effects of Sulfonamide and Tetracycline Antibiotics on Soil Microbial Activity and Microbial Biomass. Chemosphere 2005;59(4) 457–465.

159. Kotzerke A., Sharma S., Schauss K., Heuer H., Thiele–Bruhn S., Smalla K., Wilke B.M., Schloter M. Alterations in Soil Microbial Activity and N–Transformation Processes Due to Sulfadiazine Loads in Pig–Manure. Environmental Pollution 2008;153(2), 315–322.

160. Hammesfahr U., Heuer H., Manzke B., Smalla K., Thiele–Bruhn S. Impact of the Antibiotic Sulfadiazine and Pig Manure on the Microbial Community Structure in Agricultural Soils. Soil Biology and Biochemistry 2008;40(7) 1583–1591.

161. Liu F., Ying G.G., Tao R., Zhao J.L., Yang J.F., Zhao L.F. Effects of Six Selected Antibiotics on Plant Growth and Soil Microbial and Enzymatic Activities. Environmental Pollution 2009;157(5) 1636–1642.

162. Jjemba P.K. The Potential Impact of Veterinary and Human Therapeutic Agents in Manure and Biosolids on Plants Grown on Arable Land: A Review. Agriculture, Ecosystems and Environment 2002;93(1) 267–278.

163. Abu–Zreig M., Rudra R.P., Dickinson W.T. Effect of Application of Surfactants on Hydraulic Properties of Soils. Biosystems Engineering 2003;84(3) 363–372.

164. Beausse J. Selected Drugs in Solid Matrices: A Review of Environmental Determination, Occurrence and Properties of Principal Substances. Trends in Analytical Chemistry 2003;23(10) 753–761.

165. Durán–Álvarez J.C., Becerril–Bravo E., Castro V.S., Jiménez B., Gibson R. The Analysis of a Group of Acidic Pharmaceuticals, Carbamazepine, and Potential Endocrine Disrupting Compounds in Wastewater Irrigated Soils by Gas Chromatography–Mass Spectrometry. Talanta 2009;78(3) 1159–1166.

166. Fenet H., Mathieu O., Mahjoub O., Li Z., Hillaire–Buys D., Casellas C., Gomez E. Carbamazepine, Carbamazepine Epoxide and Dihydroxycarbamazepine Sorption to Soil and Occurrence in a Wastewater Reuse Site in Tunisia. Chemosphere 2009;88(1) 49–54.

167. Walker C.W., Watson J.E., Williams C. Occurrence of Carbamazepine in Soils Under Different Land Uses Receiving Wastewater. Journal of Environmental Quality 2012;41(4) 1263–1267.

168. Xu J., Chen W., Wu L., Green R., Chang A.C. Leachability of Some Emerging Contaminants in Reclaimed Municipal Wastewater-Irrigated Turf Grass Fields. Environmental Toxicology and Chemistry 2009;28(9) 1842–1850.

169. Xu J., Wu L., Chen W., Chang A.C. Simultaneous Determination of Pharmaceuticals, Endocrine Disrupting Compounds and Hormone in Soils by Gas Chromatography–Mass Spectrometry. Journal of Chromatography A 2008;1202(2) 189–195.

170. Chen F., Ying G.G., Kong L.X., Wang L., Zhao J.L., Zhou L.J., Zhang L.J. Distribution and Accumulation of Endocrine–Disrupting Chemicals and Pharmaceuticals in Wastewater Irrigated Soils in Hebei, China. Environmental Pollution, 2011;159(6) 1490–1498.

171. Pérez–Carrera E., Hansen M., León V.M., Björklund E., Krogh K.A., Halling–Sørensen B., González–Mazo E. Multiresidue Method for the Determination of 32 Human and Veterinary Pharmaceuticals in Soil and Sediment by Pressurized–Liquid Extraction and LC–MS/MS. Analytical and Bioanalytical Chemistry 2010;398(3) 1173–1184.

172. Vazquez–Roig P., Segarra R., Blasco C., Andreu V., Picó Y. Determination of Pharmaceuticals in Soils and Sediments by

Pressurized Liquid Extraction and Liquid Chromatography Tandem Mass Spectrometry. Journal of Chromatography A 2010;1217(16) 2471–2483.

173. Zeng F., Cui K., Xie Z., Wu L., Liu M., Sun G., Lin Y., Lou D., Zeng, Z. Phthalate Esters (PAEs): Emerging Organic Contaminants in Agricultural Soils in Peri–Urban Areas Around Guangzhou, China. Environmental Pollution 2008;156(2) 425–434.

174. Gielen G.J.H.P. The fate and effects of sewage–derived pharmaceuticals in soil. PhD thesis. University of Canterbury; 2007.

175. Hu X.Y., Wen B., Shan X.Q. Survey of Phthalate Pollution in Arable Soils in China. Journal of Environmental Monitoring 2003;5(4) 649–653.

176. Halden R.U. Plastics and Health Risks. Annual Review of Public Health 2010;31 179–194.

177. Ahel M., Giger W., Koch M. Behaviour of Alkylphenol Polyethoxylate Surfactants in the Aquatic Environment—I. Occurrence and Transformation in Sewage Treatment. Water Research 1994;28(5) 1131–1142.

178. Ahmad R., Kookana R.S., Alston A.M., Skjemstad J.O. The Nature of Soil Organic Matter Affects Sorption of Pesticides. 1. Relationships With Carbon Chemistry as Determined by 13C CPMAS NMR Spectroscopy. Environmental Science and Technology 2001;35(5) 878–884.

179. Chiou C.T., Kile D.E. Deviations from Sorption Linearity on Soils of Polar and Nonpolar Organic Compounds at Low Relative Concentrations. Environmental Science and Technology 1998;32(3) 338–343.

180. Hyland K.C., Dickenson E.R., Drewes J.E., Higgins C.P. Sorption of Ionized and Neutral Emerging Trace Organic Compounds onto Activated Sludge from Different Wastewater Treatment Configurations. Water Research 2012;46(6) 1958–1968.

181. Sassman S.A., Lee L.S. Sorption of Three Tetracyclines by Several Soils: Assessing the Role of pH and Cation Exchange. Environmental Science and Technology 2005;39(19) 7452–7459.

182. Durán–Álvarez J.C., Prado–Pano B., Jiménez–Cisneros B. Sorption and Desorption of Carbamazepine, Naproxen and Triclosan in a

Soil Irrigated with Raw Wastewater: Estimation of the Sorption Parameters by Considering the Initial Mass of the Compounds in the Soil. Chemosphere 2012;88(1) 84–90.

183. Karickhoff S.W., Brown D.S., Scott T.A. Sorption of Hydrophobic Pollutants on Natural Sediments. Water Research 1979;13(3) 241–248.

184. Gibson R., Durán–Álvarez J.C., Estrada K.L., Chávez A., Jiménez–Cisneros B. Accumulation and Leaching Potential of Some Pharmaceuticals and Potential Endocrine Disruptors in Soils Irrigated with Wastewater in the Tula Valley, Mexico. Chemosphere, 2010;81(11) 1437–1445.

185. Chefetz B., Mualem T., Ben–Ari J. Sorption and Mobility of Pharmaceutical Compounds in Soil Irrigated with Reclaimed Wastewater. Chemosphere 2008;73(8) 1335–1343.

186. Chefetz B., Xing B. Relative Role of Aliphatic and Aromatic Moieties as Sorption Domains for Organic Compounds: A Review. Environmental Science and Technology 2009;43(6) 1680–1688.

187. Fox K.K., Chapman L., Solbe J., Brennand V. Effect of Environmentally Relevant Concentrations of Surfactants on the Desorption or Biodegradation of Model Contaminants in Soil. Tenside, Surfactants, Detergents 1997;34(6) 436–441.

188. Gu C., Karthikeyan K.G. Sorption of the Antimicrobial Ciprofloxacin to Aluminum and Iron Hydrous Oxides. Environmental Science and Technology 2005;39(23) 9166–9173.

189. Peck A.M., Hornbuckle K.C. Synthetic Musk Fragrances in Lake Michigan. Environmental Science and Technology, 2004;38(2) 367–372.

190. Goncalves C., Dimou A., Sakkas V., Alpendurada M. F., Albanis T.A. Photolytic Degradation of Quinalphos in Natural Waters and on Soil Matrices Under Simulated Solar Irradiation. Chemosphere 2006;64(8) 1375–1382.

191. Frank M.P., Graebing P., Chib J.S. Effect of Soil Moisture and Sample Depth on Pesticide Photolysis. Journal of Agricultural and Food Chemistry 2002;50(9) 2607–2614.

192. Xia K., Jeong C.Y. Photodegradation of the Endocrine–Disrupting Chemical 4–Nonylphenol in Biosolids Applied to Soil. Journal of Environmental Quality 2004;33(4) 1568–1574.

193. Tixier C., Singer H.P., Oellers S., Müller S.R. Occurrence and Fate of Carbamazepine, Clofibric Acid, Diclofenac, Ibuprofen, Ketoprofen, and Naproxen in Surface Waters. Environmental Science and Technology 2003;37(6) 1061–1068.

194. Boreen A.L., Arnold W.A., McNeill K. Photodegradation of Pharmaceuticals in the Aquatic Environment: A Review. Aquatic Sciences 2003;65(4) 320–341.

195. Zepp R.G., Cline D.M. Rates of Direct Photolysis in Aquatic Environment. Environmental Science and Technology 1977;11(4) 359–366.

196. Latch D.E., Packer J.L., Arnold W.A., and McNeill K. Photochemical Conversion of Triclosan to 2, 8–Dichlorodibenzo–p–dioxin in Aqueous Solution. Journal of Photochemistry and Photobiology A: Chemistry, 2003;158(1) 63–66.

197. Calisto V., Domingues M.R.M., Erny G.L., Esteves V.I. Direct Photodegradation of Carbamazepine Followed by Micellar Electrokinetic Chromatography and Mass Spectrometry. Water Research 2011;45(3) 1095–1104.

198. Ying G.G., Yu X.Y., Kookana R.S. Biological Degradation of Triclocarban and Triclosan in a Soil Under Aerobic and Anaerobic Conditions and Comparison with Environmental Fate Modeling. Environmental Pollution 2007;150(3) 300–305.

199. Gasser G., Rona M., Voloshenko A., Shelkov R., Tal N., Pankratov I., Elhanany S Lev O. Quantitative Evaluation of Tracers for Quantification of Wastewater Contamination of Potable Water Sources. Environmental Science and Technology, 2010;44(10) 3919–3925.

200. Li J., Dodgen L., Ye Q., Gan J. Degradation Kinetics and Metabolites of Carbamazepine in Soil. Environmental Science and Technology 2013;47(8) 3678–3684.

201. Walters E., McClellan K., Halden R.U. Occurrence and Loss Over Three Years of 72 Pharmaceuticals and Personal Care Products from Biosolids–Soil Mixtures in Outdoor Mesocosms. Water Research 2010;44(20) 6011–6020.

202. Kormos J.L., Schulz M., Kohler H.P.E., Ternes T.A. Biotransformation of Selected Iodinated X–ray Contrast Media and Characterization of Microbial Transformation Pathways. Environmental Science and Technology 2010;44(13), 4998–5007.

203. Pomiès M., Choubert J.M., Wisniewski C., Coquery M. Modelling of Micropollutant Removal in Biological Wastewater Treatments: A Review. Science of the Total Environment 2013;443 733–748.

204. Kookana R.S., Ying G.G., Waller N.L. Triclosan: its Occurrence, Fate and Effects in the Australian Environment. Water Science and Technology 2011;63 (4): 598–604.

205. Schlüsener M.P., Bester K. Persistence of Antibiotics such as Macrolides, Tiamulin and Salinomycin in Soil. Environmental Pollution, 2006;143(3) 565–571.

206. Xu G., Li F., Wang Q. Occurrence and Degradation Characteristics of Dibutyl Phthalate (DBP) and Di–(2–ethylhexyl) Phthalate (DEHP) in Typical Agricultural Soils of China. Science of the Total Environment 2008;393(2) 333–340.

207. Marco–Urrea E., Pérez–Trujillo M., Blánquez P., Vicent T., Caminal G. Biodegradation of the Analgesic Naproxen by Trametes versicolor and Identification of Intermediates Using HPLC–DAD–MS and NMR. Bioresource Technology 2010;101(7) 2159–2166.

208. Gauthier H., Yargeau V., Cooper D.G. Biodegradation of Pharmaceuticals by Rhodococcus rhodochrous and Aspergillus niger by co–Metabolism. Science of the Total Environment 2010;408(7) 1701–1706.

209. Chao W.L., Lin C.M., Shiung I.I., Kuo Y.L. Degradation of Di–butyl–phthalate by Soil Bacteria. Chemosphere 2006;63(8) 1377–1383.

210. Jelic A., Cruz–Morató C., Marco–Urrea E., Sarrà M., Perez S., Vicent T., Petrovic M., Barcelo D. Degradation of Carbamazepine by Trametes versicolor in an Air Pulsed Fluidized Bed Bioreactor and Identification of Intermediates. Water Research 2012;46(4) 955–964.

211. Shenker M., Harush D., Ben–Ari J., Chefetz B. (2011). Uptake of Carbamazepine by Cucumber Plants–A Case Study Related to Irrigation with Reclaimed Wastewater. Chemosphere, 2011;2(6) 905–910.

212. Wu C., Spongberg A.L., Witter J.D., Fang M., Czajkowski K.P. Uptake of Pharmaceutical and Personal Care Products by Soybean Plants from Soils Applied with Biosolids and Irrigated with

Contaminated Water. Environmental Science and Technology 2010;44(16) 6157–6161.

213. Limousin G., Gaudet J.P., Charlet L., Szenknect S., Barthes V., Krimissa M. Sorption Isotherms: a Review on Physical Bases, Modeling and Measurement. Applied Geochemistry 2007;22(2) 249–275.

214. Murillo–Torres R., Durán–Álvarez J.C., Prado B., Jiménez–Cisneros B.E. Sorption and Mobility of Two Micropollutants in Three Agricultural Soils: A Comparative Analysis of Their Behavior in Batch and Column Experiments. Geoderma 2012;189–190 462–468.

215. Bi E., Schmidt T.C., Haderlein S.B. Environmental Factors Influencing Sorption of Heterocyclic Aromatic Compounds to Soil. Environmental Science and Technology 2007;41(9) 3172–3178.

216. Maoz A., Chefetz B. Sorption of the Pharmaceuticals Carbamazepine and Naproxen to Dissolved Organic Matter: Role of Structural Fractions. Water Research 2010;44(3) 981–989.

217. Stumpe B., Marschner B. Long–Term Sewage Sludge Application and Wastewater Irrigation on the Mineralization and Sorption of 17β–Estradiol and Testosterone in Soils. Science of the Total Environment 2007;374(2) 282–291.

218. Kaiser K., Guggenberger G., Zech W. Sorption of DOM and DOM Fractions to Forest Soils. Geoderma 1996;74(3) 281–303.

Characterization and Remediation of Soils and Sediments Polluted with Mercury: Occurrence, Transformations, Environmental Considerations and San Joaquin's Sierra Gorda Case

Robles[1], J. Lakatos[2], P. Scharek[3], Z. Planck[3], G. Hernández[4], S. Solís4, and E. Bustos[1]

[1]Centro de Investigación y Desarrollo Tecnológico en Electroquímica, S. C. Parque Tecnológico Querétaro S/N, Sanfandila, Pedro Escobedo, Querétaro, Mexico

²University of Miskolc, Faculty of Material Sciences and Engineering, Institute of Chemistry, Miskolc, Hungary

³Geological and Geophysical Institute of Hungary, Budapest, Hungary

⁴Centro de Geociencias, Campus UNAM-Juriquilla, Querétaro, Qro., Mexico

INTRODUCTION

Soil as important part of the ecosystems which must be protected in the environment context, and it is necessary be studied the possible overall impact of measures for protection, with a very special attention from mining activities. The soil resource occupies a fundamental part of the ecosystems; when a soil is degraded, the others components of the ecosystems are degraded too.

The fate of the heavy metal in soils depends upon many soil processes that are governed by several soils properties of which soil pH and redox potential are known to be the most important parameters. Thus, the solubility of trace elements is often shown as a function of pH affected by amount and kind of organic matter. Trace elements are known to be accumulated in surface soils as a result of contamination from point sources as mining activities. An appreciable amount of the soils has been made unusable because of pollution. Highly contaminated soils belong to a high healthy risk to human being and their environmentally harmful effects. That is why soil should be correctly understood and underestimated long range lethal effects that can have irreversible consequences. The improvement of soils damaged and contaminated by pollutants need of the particular soils, requires a full understanding of soil properties and of the deteriorating factors.

Mercury is one of the most toxic elements to human health and ecosystem; because of all mercury species are toxic. A wide variety of mercury species exist in the environment and its various chemical forms can differ in bioavailability, transport, persistence, and toxicity. Still, every mercury species is toxic with methyl mercury being the most toxic species. The World Health Organization (WHO) recommends a maximum methyl mercury intake of 1.6 μg Kg⁻¹ per week, while the Environmental Protection Agency (EPA) lists a maximum recommended intake of 0.1 μg Kg⁻¹ of body weight per day for adults. Due to high bioaccumulation, mercury is found on many levels of the food chain (Hinton and Veiga, 2001; Bengtsson, 2008). Any form of mercury in the environment may evolve into a more toxic species (methyl mercury) under biogeochemical transformation processes (Figure 1). Due to these processes and the high mobility of mercury species, a good understanding of how mercury species transform and accurate monitoring are essential for assessing the risk of mercury in the environment.

The impact of mercury depends strongly on its chemical species; understanding mercury transformations and the impact of its various chemical forms are vital to preventing harmful effects on humans and the environment. Nevertheless, the physicochemical characteristics of mercury are either useful or necessary for many industrial and agricultural applications, and mercury may be scattered over large area, depending on the source (Leopold et al, 2010).

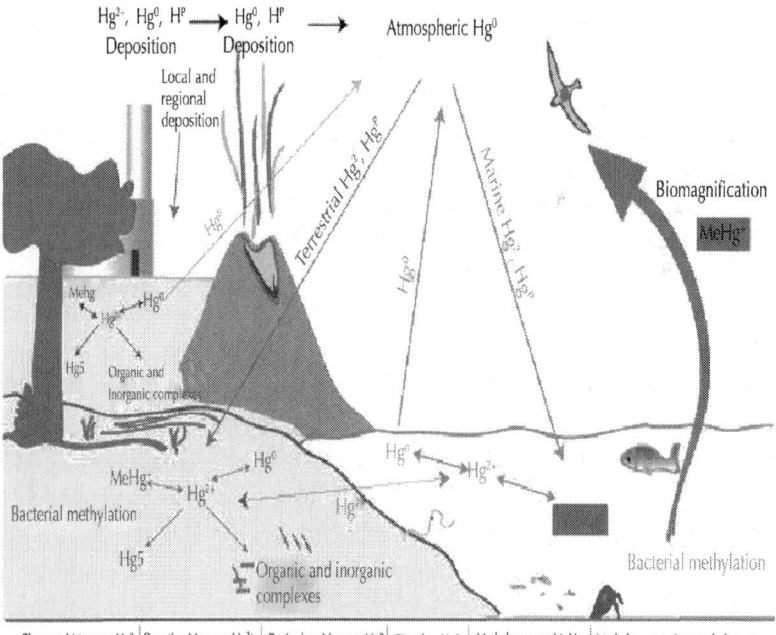

Figure 1: Biogeochemical transformation processes of mercury (Leopold et al, 2010).

Mercury concentrations in ground water indicate that the highest concentration of mercury in groundwater comes from the soil and from aquifers. While simulating mercury predictions can often be difficult, mercury can be estimated in experiments conducted in batch mode or in columns. The percentage of Hg that can potentially leach from the soil was previously estimated in batch experiments. Distribution analyses of species in leachate confirmed the presence of inorganic species (Hg^{2+} and Hg^0) ranging from 90 – 100 % (Bollen, 2008; Harvey, 2002).

The most commonly used techniques for the remediation of mercury contaminated soils have been classified as either excavation techniques or containment techniques, and are grouped as follows (Hinton and Veiga, 2001): (a) *ex situ* treatments: physical separation, thermal treatments, hydrometallurgical treatments; (b) *In situ* recuperation: vapor extraction coupled with evaporation (soil), permeable reactive barriers; (c) *In situ* leaching and extraction: electrokinetic separation, interceptor systems, phytoremediation, passive remediation; (d) *containment:* pump and treat impermeable barriers, sealed surfaces and drainage, stabilization and solidification, sediment covering.

PHYSICOCHEMICAL PROPERTIES OF MERCURY

Mercury (Hg) is a chemical element with an atomic number of 80. Mercury is a silver plated heavy metal, liquid and odorless at normal conditions. It easily alloys with many other metals like gold or silver producing amalgams, is insoluble in water and soluble in nitric acid. The main source of Hg is cinnabar or mercury sulfide (HgS), a stable compound and insoluble usually recovered as a byproduct of ore processing. Mercury in this form is found in the earth's crust average concentrations of 0.5 ppm (Hinton and Veiga, 2001).

Mercury is one of the most toxic elements to human health and ecosystem. At temperatures above 40 °C mercury produces toxic and corrosive fumes. It is harmful by inhalation, ingestion and contact, is a very irritating to skin, eyes and respiratory tract, even to nervous system, its gaseous form is absorbed by lung tissues (Hinton and Veiga, 2001, Bengtsson, 2008).

A wide variety of mercury species exist in the environment and its various chemical forms can differ in bioavailability, transport, persistence, and toxicity. Still, every mercury species is toxic with methyl mercury being the most toxic species. This element can exist in the environment as elemental (Hg^0), oxidized inorganic (Hg^{2+} -mercuric, Hg_2^{2+}-mercurous) or oxidized organic (methyl/ethyl mercury) forms. Mercuric and mercurous forms are more stable under oxidizing conditions. In moderately reducing conditions, the organic or inorganic mercury can be reduced to its elemental form and be converted to forms leased by biotic or abiotic processes: these are the most toxic forms of mercury, as well as being soluble and volatile. Hg (II) forms strong soluble complexes with a variety of organic and inorganic ligands oxidized in aqueous systems. Hg sorption in soil, sediment and humic materials is an important mechanism for the removal of mercury from solutions, another mechanism, a high pH is their co-precipitation sulfide (HgS) (Leopold et al, 2010).

Any form of mercury in the environment may evolve into a more toxic species (methyl mercury) under biogeochemical transformation processes. Due to these processes and the high mobility of mercury species, a good understanding of how mercury species transform and accurate monitoring are essential for assessing the risk of mercury in the environment. The impact of mercury depends strongly on its chemical species; understanding mercury transformations and the impact of its various chemical forms are vital to preventing harmful effects on humans and the environment. Nevertheless, physicochemical characteristics of mercury are either useful or necessary for many industrial and agricultural applications, and mercury may be scattered over large area, depending on the source (Leopold et al, 2010; Nick, 2012).

The metal mercury (Hg^0) is mainly used to produce chlorine gas and caustic soda, and is part of some types of alkaline batteries, fluorescent lamps, electrical contacts, and instruments such as pressure gauges and thermometers, among others. Hg salts are used in antiseptic ointments and creams and skin lightening. Among the activities that generate the most pollution by Hg, is the burning of coal and chlor-alkali plants: other important sources are mining and metallurgy and the burning of municipal solid waste, which may contain instruments such as pressure gauges, thermometers, alkaline batteries and fluorescent lamps. The mercury released into the air tends to settle and adhere to soil organic matter (Hinton and Veiga, 2001; Nick, 2012).

Natural and anthropogenic mercury emissions are mainly in the form of elemental mercury (Hg^0), which makes up about 99 % of total atmospheric mercury. However, biogeochemical transformations can oxidize it, forming Hg^+ and Hg^{2+}. Most inorganic Hg compounds are water soluble in small doses, and can be found in soil and sediments. In contrast, the presence of inorganic forms of Hg^{2+} bonded to organic and/or inorganic species ($[HgCl_x]^{2-x}$; $[Hg^{II}\text{-}DOC]$; $[HgS]$) depends on the local chemical environment. The life time of these compounds in air is very short (on the scale of minutes) and they are rapidly removed by deposition processes because of high water solubility and surface activity. Figure 2 shows the main mercury species in the atmosphere, hydrosphere and sediment (Leopold et al, 2010; Nik, 2012; Slowey et al, 2005; Wartel et al, 1999; Shi et al, 2005).

Over 90 % of surface water mercury is from atmospheric deposition. Hg^{2+} usually undergoes a biomethylation process that forms methylmercury (MeHg, CH_3Hg^+) and dimethyl mercury (DMeHg, $(CH_3)_2Hg$), though these reactions can be reversed using microorganisms and/or photolytic decomposition. All these species are highly mobile.

Three main forms of mercury are found in natural waters: elemental mercury (Hg^0), inorganic Hg^{2+} (Hg^{2+} and its complexes) and organic mercury (MeHg, MeHg complexes and DMeHg). With solubility (at 25 °C) of 0.08 mg L^{-1}, Hg^0 can be found at all depths. Inorganic mercury (Hg^{2+}) and MeHg forms complexes with other dissolved compounds in fresh water, but for the most part, only forms complexes with chlorine in sea water. DMeHg is found in the deep sea. (Wartel et al, 1999, Shi et al, 2005, Slowey et al, 2005).

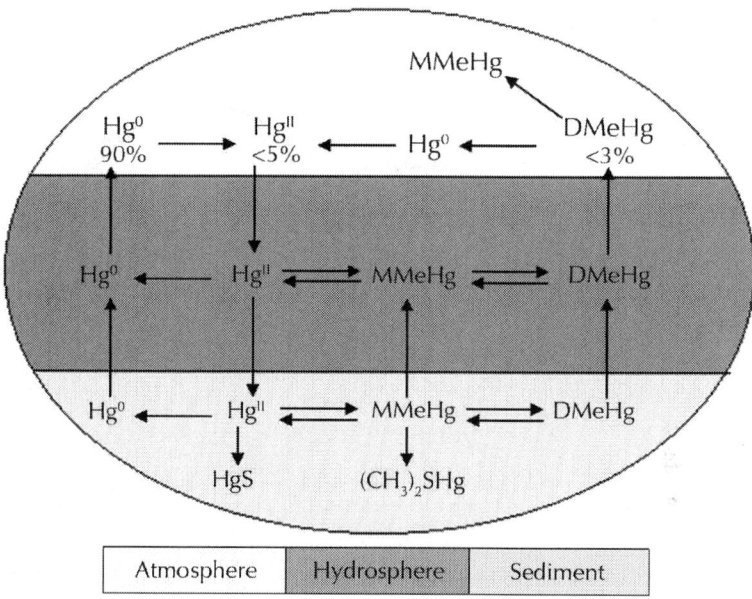

Figure 2: Distribution of mercury species in atmosphere, hydrosphere and sediment (Leopold et al, 2010).

EDAPHOLOGY PROPERTIES OF SOIL RELATED WITH MERCURY

Soil is a collection of natural bodies on the Earth' surface, in places that most of them have been modified by man in its quality and containing living matter and supporting or capable of supporting plants. Soil grades at its lower margin to hard rock or to earthy materials virtually devoid of roots, animals or marks of other biologic activity.

In general, physical, chemical and biological soil characteristics are highly correlated parameters that are necessary to understand. The specific elemental composition of each particular soil reflects, to a degree modified over time by weathering and the chemical composition of the parent material from which the soil is formed. For instance, the extractability of the different elements depends on the soil properties.

Soil as important part of the ecosystems which must be protected in the environment context, and it is necessary be studied the possible overall impact of measures for protection, with a very special attention from mining activities. The soil resource occupies a fundamental part of the ecosystems; when a soil is degraded, the others components of the ecosystems are degraded too.

The fate of the heavy metal in soils depends upon many soil processes that are governed by several soils properties of which soil pH and redox potential are known to be the most important parameters. Thus, the solubility of trace elements is often shown as a function of pH affected by amount and kind of organic matter. Trace elements are known to be accumulated in surface soils as a result of contamination from point sources as mining activities.

An appreciable amount of the soils has been made unusable because of pollution. Highly contaminated soils belong to a high healthy risk to human being and their environmentally harmful effects. That is why soil should be correctly understood and underestimated long range lethal effects that can have irreversible consequences. The improvement of soils damaged and contaminated by pollutants need of the particular soils, requires a full understanding of soil properties and of the deteriorating factors.

Mercury is a microelement: its Clark value in the Earth's crust is 56 $\mu g\ Kg^{-1}$ (Fügedi et al, 2011). It is characterized by a dual geochemical behavior: it is liable to extreme concentration and to dispersion, the latter resulting in an approximately entirely even concentration. It is found either as a native metal (near to 80 % in hydrothermal and vapors) or in cinnabar, corderoite, livingstonite and other minerals. Cinnabar (HgS) is the most common ore. Mercury ores usually occur in very young orogenic belts where rock of high density on upper mantle is forced to the crust of the Earth (Ozerova, 1996).

Given that mercury is enriched by an extremely wide variety of geological processes (Fergusson, 1990) from the formation of hydrocarbon to hydrothermal mineral occurrences, it can be regarded as an universal geochemical indicator of young geological effects; its dispersion halos are more extensive than that of any other element (Fügedi et al, 2011).

Historically there were two main registered Mercury mines: Almaden (Spain) and Idrija (Slovenia) in Europe. Later new occurrences were found in California and worldwide. It was used in gold separation. In Mexico we know mercury mines from the Pre-Hispanic era (Scharek et al., 2010) and a usage in cultic fests (Figure 3). In 2005, China was the top producer of mercury with almost two-thirds global share followed by Kyrgyzstan. Several other countries are believed to have unrecorded production of mercury from copper electro winning processes and by recovery from effluents.

Figure 3: Typical cinnabar occurrences in a limestone system (Formation Las Trancas, San Joaquin, Querétaro, Mexico, photo by P. Scharek).

MOBILIZATION AND LOADINGS THE MOBILIZED MERCURY

The mobilization of the different mercury forms can due to by evaporation and dissolution. Here we do not deal with the erosion which able to mobilize all form of mercury if it has attached to the solid particle, however in case of soils and fly as particles in the flue gas this particle associated mobilization mechanism can play important role in the mercury transport.

Evaporation

The evaporation governed by the vapor pressure depends on the volatility of the compound and the temperature. Concerning the volatility of the different mercury compounds evaporation at ambient temperature can be significant in case of elemental and the organic mercury cases, however the volatilization of the other mercury forms (the inorganic mercury compounds) can become considerable if the temperature reaches a couple hundred degree centigrade. All of the mercury compounds have relatively low boiling points (Table 1), some of them decompose before melting, others can sublimate. Based on these data it is obvious the vaporization can play important role of the mercury compound transport and mobilization.

Table 1: Melting and boiling points of the mercury and mercury compounds

Hg	Melting T, °C	Boiling T, °C
	-38.9	356.5
HgS (cinnabar, α)	580 sublimate	
HgS (metacinnabar, β)	446 sublimate	
Hg2O	100 decompose	
HgO	500 decompose	
Hg SO4	450 decompose	
Hg2Cl2	302	384
HgCl2	277	304
HgBr2	237	322

HgI2	259	354
CH3HgCl	-	92
(CH3)2Hg	-43	94

The partial pressure of the elemental mercury (Hg) reach 1 Pa at 42 °C and enhances exponentially till the boiling point (T_b = 356.5 °C). At 20 °C the Hg vapor pressure is 0.18 Pa the Hg concentration in the air saturated with the mercury is 7.64 10^{-8}mol dm^{-3} = 15.3 µg m^{-3}. Due to this high volatility the elemental mercury evaporate if stored and processed an open container. Elemental mercury can escape from solution if the oxidized mercury is able to reduce. The analytical data will be inaccurate if the sample is not preserved agents the elemental mercury formation.

Concerning the global mercury contamination till the middle of the past century evaporation of elemental mercury used to extract silver and gold was the main source of the mercury emission, (Nriahu, 1994). All the once produced and recently available elemental mercury stock (in the past five century one million tons was produced from cinnabar and from other ores) if in used either evaporation or after transformation can contribute to the mercury contamination worldwide (Hylanderand Meili, 2003). This is the reason why the elemental mercury use is banned. Recently one of the most significant sources of mercury emission by evaporation is the coal firing. During the coal burning the mercury associated with pyrite and be organically bonded to the coal minerals are released in the combustion flame as elemental mercury, which is partially oxidized to Hg(II) in homogeneous and heterogeneous catalytic reaction governed by the chlorine and the ash content of the combustion gases (Sondreal et al, 2004).

Generally accepted view is that the evaporated oxidized forms of mercury contaminate the environment locally, close to the emission source. However they can transform to elemental mercury and depend on this transformation rate it can become part of the global mercury cycle.

The transformation of the oxidized to reduce the reduced to oxidized forms can happen both in gas and aquatic environment according to the circumstances. There is similar transformation between the inorganic and organic forms.

Different species of macro algae from the dissolved mercury can produce different methylated mercury compounds in the ocean. Because of these methylated mercury compounds have high volatility and at the dimethylated form has low solubility in ocean water they are easily emitted into the atmosphere and can contribute significantly to the global atmospheric mercury (Pongratz and Heuman, 1998). Beside this different bacteria (e.g. sulfate reducing) and in case of abiotic route the tin- alkyls and the humic acids also can transform the dissolved mercury (II) to methyl mercury form (Weber, 1993). A quite detailed set of possible transformation in gas and aquatic media and the Henry constants which inform about the dissolved compound volatility are collected by Shon at al, 2005.

During heating mercury compounds can transforms directly or via oxides to elemental mercury. Beside the elemental mercury only the halogenides and the sulfides since last have a tendency to sublimate can occur in evaporated forms. The sulfides at presence of oxygen at 600 °C transforms to Hg and SO_2, however in presence of Fe and CaO the HgS also will decompose to Hg and Fe- or Ca- sulfides. Using the temperature programmed evaporation technique based on the volatility difference of mercury compounds the compound forms can be distinguished and can use for mercury speciation in solids (Lopez-Anton et al, 2010, 2011).

In a high temperature process since the mercury compounds decompose the original speciation of mercury does not preserve a new speciation can formed which determined by the gas composition. In the high temperature gases high portion of mercury exists in elemental and just a small portion in oxidized form. This is the reason why these technologies such as coal fired energy production, the cement kiln, the incineration has difficulty in the mercury capture.

Focusing to the soil, the heating comes from sunlight can mobilize only the weakly sorbed elemental and organic mercury but the fire on the soil surface, for example the forest fire can evaporate the less volatile mercury forms as well. This case the contamination level of the fired soil decreases but, due to the transport, at other places the contamination becomes higher (Caldwel et al, 2000).

The volatilization can be the cause of contamination but can use for decontamination as well. Based on the volatilization of mercury compounds, mercury removal process was established from coal cleaning by mild pyrolysis (Wang et al., 2000) and for the soil cleaning by thermal treatment. In case of coals the speciation of mercury determines the maximum efficiency of the mercury removal. The efficiency of the process generally remains below 100 %, (bituminous coal case at 500 °C it was aprox. 75 %). Since the efficiency remains below 100 % the rest of mercury still remain in the process and pass to the flue gas after the coal burning. The speciation of the mercury in the contaminated soil also has influence on the efficiency of the thermal remediation, see more details later.

Concerning that the different mercury forms exhibit different volatility the actual distribution of the mercury species in a medium the rate of the transformation process which able to modify it together govern the mercury mobilization by evaporation.

It is well known, if elemental mercury forms in the water this elemental mercury can easily escape to the gas phase. It is quite intensive if gas bubbling through the water or the water surface is disturbed (Okouchi and Saaski, 1984). Sunlight induced H_2O_2 formation in alkaline condition can result reduction of the oxidized mercury forms to elemental mercury. This can explains that the Hg concentration above the lake water surface can be higher day time than night. The fulvic and humic compounds are able to complex the mercury (II) ion in aquatic media but these compounds can take part in the mercury alkylations, further at a suitable pH can work as a reducing agent. The redox potential at 0 pH for Hg (II) reduction is 0.85 V(Allard and Arsenie, 1991).

This type of mercury transformation between oxidized and reduced forms together with the alkylation will generate not only a modification between the concentrations of the mercury species in the aquatic phase but will modify the mercury transport between the phases. The mercury transformation processes are important in the technological processes used for the mercury removal since can effect they efficiency (Somoano et al., 2007).

Dissolution

The mobilization by dissolution can arrange two groups: (a) dissolutions ways exist in the nature (b) dissolution way can be applied in the laboratory and in the remediation technology to determine the loading forms or remove the mercury from the contaminated media.

Dissolutions Ways Exist in the Nature

The solubility of elemental mercury and the ore of mercury can find in the nature (cinnabar etc.) are very low in water. This low solubility result low mercury concentration level in aquatic phase and restricts the transport between phases by dissolution. However the oxidation both cases enhances these mercury forms solubility. The elemental mercury can be oxidized by ozone, halogens, some components of acid rains, or by oxy-acids in laboratory (HNO_3 and the hot H_2SO_4) resulting a soluble form. The ozone in air if does not consumed by the other more reactive air contaminants can oxidize Hg to Hg(II) (Iverfeld and Linquist, 1986; Shonet al, 2005).

In aquatic media oxidation can occur at acidic conditions if the sunlight produces oxidative radicals OH, or peroxides. This process can play role in the trap of the physically dissolved elemental mercury in water, and also can hinder the transformation of the oxidized mercury forms towards the reduced elemental mercury direction. The oxidative transformation of elemental mercury is essential in case of many mercury capture process since the oxidized forms of mercury has higher tendency to sorb and dissolve, therefore different oxidation procedures are available and applied in the demercurysation technologies (Ko et al, 2008; Lakatos et al, 2009; Sondreal et al, 2004).

However the mercury in the natural minerals is in the oxidized forms these minerals luckily due to the very low solubility can be considered not a mobile occurrence of the mercury. The environmental risk improves if the natural processes can transform the minerals a more soluble form. One of the most significant ore transformations which effect the mercury mobilization is the sulfide ore oxidation in the air. The oxidations of sulfides to sulfate a considerable enhancement ensue in mercury solubility (Holley et al, 2007). This process, the oxidation of the tailings, can accused for the mercury contamination all around the abandoned mercury ore mines.

Table 2: Solubility of different mercury compounds

Compound	Solubility in Water c, ppm

Hg	0.049*
HgS (cinnabar,)	0.01
HgS (metacinnabar,)	-
HgSO4	-
Hg(NO3)2	soluble
Hg2O	51
HgO	51
Hg2Cl2	10
HgCl2	66 000*
HgBr2	5 100
HgI2	51
CH3HgCl	5 780
(CH3)2Hg	-

[i] - *Solubility from paper of Ko et al, 2008.

Among the mercury compounds (Table 2) the mercury-chloride and nitrates are those which have the highest solubility in water. The simple cationic form of Hg(II) is not the common form in the aquatic media, it exist only in acidic solutions, at less acidic condition the dissolved mercury appears as $HgOH^+$, HgOHCl, $Hg(OH)_2$ and $HgCl_2$ molecules and complex anions $HgCl_4^{2-}$ at high chloride concentration. It means that the sea water contains the oxidized mercury mainly in this chlor-complex form. Beside the chlor- complex the mercury -fulvo and -humic complexes also exists in aquatic environment. The speciation in the solution, the molecular forms govern the mercury loadings and play important role at the way and efficiency of the removal.

Dissolution for Leaching Mercury from Different Medium

Beside the thermal way and the application of the species sensitive analytical methods for mercury analysis (XPS, EXAFS etc.) the sequential extraction is often applied technique to specify the mercury chemical form and associations in solids. The thermal methods and the species sensitive elemental analysis can distinguish the elemental Hg, the HgS forms and the organically bonded mercury forms which generally exist in the soil. However these techniques do not allow doing any estimation about mobility and bioavailability of mercury. To get the loading specific information in soil for mercury, beside the classical Tessier six step extraction used generally, different modified procedure are available for Hg which able to distinguish better the mercury forms than the Tesssier method can do (Orecchio and Polizzotto, 2013; Han et al,2006). For example it can determine mercury bounded to amorphous iron oxides (by NH_4 oxalate-oxalic acid extraction), mercury bonded to crystalline iron oxides (by $NH_2OH \bullet HCl$- 25 % acetic acid extraction), non-cinnabar mercury (elemental mercury, organic bounded, humine bounded (by 4 M HNO_3 extraction), cinnabar mercury (by extracted with saturated Na_2S,Han et al, 2006). The advantage of this protocol is the ability of the separation of humic and sulfide bounded mercury which important in the soil case. Two set of sequential extraction regime can compared at Table 3 and 4.

Table 3: Mobilization protocol for determination of association of mercury to soil component (Han et al, 2006)

Mobilization Protocol	Determination of Association of Mercury to Soil Component
Loading form of mercury	Extraction procedure
Water soluble mercury	NH4-acetate
Exchangable mercury	(1 M NH4-acetate pH 7 set with NH4OH: solid:liquid 1:25, 30 min 25 °C)
Carbonate bounded mercury	Hidroxylamine – HCl
Easily reducible oxides bounded mercury (Mn-oxides)	(0.1 M NH2OH.HCl +0.01 M HCl solid:liquid1:25, 30 min 25 °C)
Elemental and organic bounded mercury	H2O2 (3 mL 0.1 M HNO3+5 mL 30 % H2O2 80 °C 2 h; 2 mL H2O280 °C 1 h; 50 mL 1M NH4- acetate)
Amorphous iron-oxide bounded mercury	NH4-oxalate –oxalic acid (0.2 M oxalate buffer 1:1, pH 3.25 solid:liquid1:25)

Crystalline iron oxide bounded mercury	Hydroxilamine – HCl- acetic acid (hot) (0.04 M NH2OH.HCl in 25 % acetic acid 97-100 °C 3 h solid:liquid 1:25)
Non cinnabar bounded mercury (Hg, organically bounded, humin bounded) TOT (non cinnabar mercury)	4 M HNO3 (4 M HNO3 80 °C16 h solid:liquid1:25)
Cinnabar bounded mercury	Na2S (4 mL saturated Na2S 12 h repeated twice)

Table 4: Mobilization protocol for determination of association of mercury to soil component (Orecchio and Polizzotto, 2013)

Mobilization Protocol	Determination of Association of Mercury to Soil Component
Loading form of mercury	Extraction procedure
Water soluble mercury	H2O100°C, 1h stirred, solid:liquid 1:8
Exchangable mercury	Na-acetate (1M Na-acetate, 1 h stirred, solid:liquid1:8
Carbonate bounded mercury	Na-acetate –aceticacidpH 5 (1M Na-acetate –acetic acid pH 5, 4 h stirred, solid:liquid1:8)
Fe, Mn-oxide bounded mercury	Hydroxylamin HCl - acetic acid 0,04 M NH2OH•HCl in 25 % Acetic acid, 96 °C, 6 h, Solid:liquid 1:8
Elemental and organic bounded mercury	Mineralisation by HNO3-H2O2 (a)heated previously 180 °C - organic bounded - microwave digestion in cc HNO3-H2O2 mixture; (b) no heat - elemental + organic bounded - microwave digestion in cc HNO3-H2O2mixture)
Sulfid bounded mercury	Aqua regia (HCl:HNO3 3:1)

Mobilization of mercury can occur through complex formation, ligand exchange reactions with chloride and sulfur-containing ligands which leading to enhanced Hg solubility in soil solutions. The sulfur containing ligands: tiosulfates ($S_2O_3^{2}$), thyocyanites (SCN^-) can mobilize the mercury efficiently and could improve the phytoextraction efficiency (Moreno et al, 2004).

Removal mercury by phytoextraction from soils and others soil like materials eg. waste water plants biosolids often need additives which improve the solubility of the mercury. These mobilizations agents are used in accelerate phytoextraction. One type is the chelating agents: citrate, oxalate, malate, succinate, tartarate, salicilate, acetate, and amino-poly-carboxylic acids: EDTA (Lomonte et al, 2011). Since the EDTA is persistent compound, recently the biodegradable ethylendiamine-disuccinate (EDDS) or nitrilotriacetic acid (NTA) suggested as alternative chelator instead of EDTA (Evangelu et al, 2007).

Specific compounds used the mercury extraction from tissues: they can pay role in case of poisoning for detoxification : EDTA was tested for detoxification by Aposhian (Aposhian et al,1995), 2,3- dimercapto-1- propansulfonate was used to extract mercury from tissues of rats exposed to different mercury compounds (Buchet and Lauwerys, 1989). The EDTA was not found the best for mercury removal in human application since does not the best chelator for Hg and has side effects, however there are two other compounds which suggested to keep in stock in any poison control center as mercury chelator DMPS (2,3-dimercapto-1-propane sulfonic acid, (unithiol)) and DMSA (meso-2,3 dimercapto succinic acid, succimer, Guzzi et al, 2010).

Loadings of Mercury

The mercury contaminations are in the environment can exist different phases: as vapor in the air (Hg, and compounds, particle associated), dissolved in aquatic media (Hg^{2+}, $Hg(OH)_2$, $HgCl_2$, $HgCl_4^{2-}$, different complexes, particle associated) and solid as precipitates or minerals and in associated forms bonded to different manner to the component of different solids (soil, fly ash, waste water sludge, etc.). In the previous section it was demonstrated that how the associations can be identified.

The loadings of mercury to solid can be considered as positive or negative phenomenon. It can restrict the dispersion of the contamination one side it is positive, the negative this way it can preserve the contamination. Since the loadings depends on the character of the collector and the speciation of the mercury, difficult to establish general rules for this process. However it can state that the elemental mercury has low sorption ability, the cationic sorbs better than the anionic forms on clays, and negatively charged carbon surfaces (coals, activated carbons, humic materials) the loading is more effective to that surfaces which have contain sulfides. It was interesting findings after the cinnabar oxidation a part of liberated mercury could load to the cinnabar surface this way it can be not just the source but the collector of mercury ions. Unfortunately the most toxic forms (alkyls) have the highest ability for bioaccumulation.

The nature works against the mercury contamination. Except the alkylation it transforms the mercury toward the most stable less soluble form. Near the chlor-alkali plant the total mercury sometimes reach the four order of magnitude higher level, than the background concentration, luckily it found a non-volatile and non-soluble associations since transforms to sulfides (Bernaus et al, 2006).

The history of mercury contamination is recorded by loadings. The dept profile of mercury concentration on peat can provide a clear picture how the mercury contamination changed during the mankind history (Barraclough et al, 2002).

The loadings play important role in the environmental technologies used for decrease the mercury emission or clean the contaminated medium. Sulfur and halogen containing carbons, oxidative inorganic sorbents (Lakatos et al, 2009) were developed for elemental mercury removal from flue gas. Beside a range of, classical, functionalized sorbents, sulphur containing carbon nanotubes widen the collection one can chose among for eliminate the mercury contamination in aquatic media (Pillay et al, 2013).

The coals especially the low rank and the oxidized coals are very good mercury ion collectors. Due to this feature we must face that the coal-firing are the main source of the anthropogenic mercury contamination nowadays. However this material offers us an application for cure a slice of the mercury problem: remove the mercury from aquatic media. It can use in batch mode or dynamic systems as the reactive barrier material, by the high mercury capture coal are able to retard or remove the aquatic mercury contamination (Lakatos et al, 1999).

REMEDIATION OF POLLUTED SOIL WITH MERCURY

The most commonly used techniques for the remediation of mercury contaminated soils have been classified as either excavation techniques or containment techniques, and are grouped as follows (Hinton and Veiga, 2001): excavation and *ex-situ* treatments, containment and *in situ* chemical treatment (Figure 4).

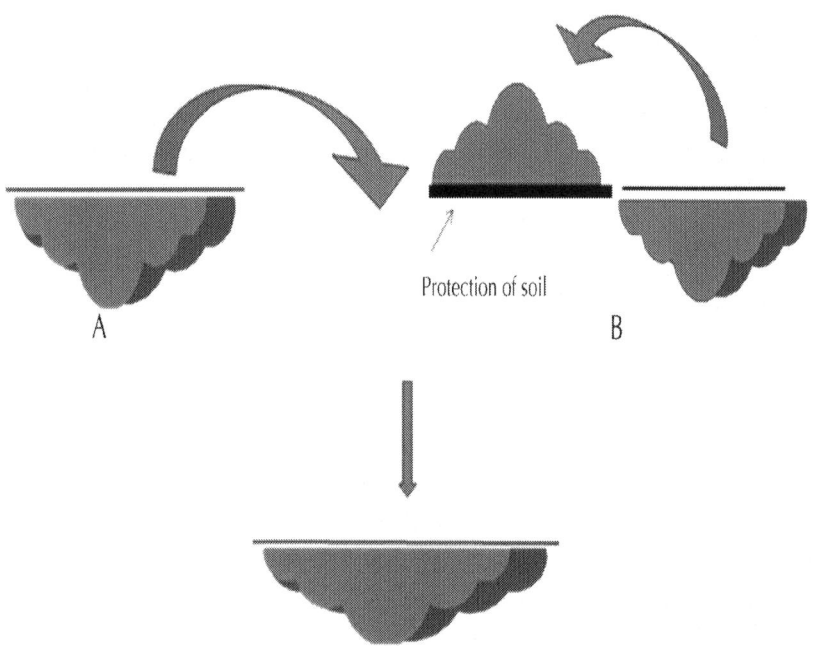

Protection of soil

A   B

Figure 4: Representation of the techniques for the remediation of mercury contaminated soils: (A) excavation and *ex-situ* treatments, (B) containment and (C) *in situ* chemical treatment.

Excavation and *Ex Situ* Treatments

They will be treated off-site soil contaminated when removing soil or contaminated soil-like materials to a place outside the place in which they are located, for submission to authorized treatment fixtures. (Hinton and Veiga, 2001):

- *Physical separation.* Mercury has affinity for the smallest particles of soil.

- *Thermal treatment.* The volatility of mercury increases with increasing temperature; therefore, a heat treatment technique of excavated soil is a potentially effective technique for the removal of mercury in soils.

- *Hydrometallugical treatment.* Chemical extraction of mercury in contaminated soils can be induced by four mechanisms: desorption of adsorbed species, oxidation of metallic mercury, use of strong complexing agents, and dissolution of Hg precipitate. The efficiency and mechanism employed decrease with respect to time due to recomplexation, readsorption and removal of the soluble fraction. Two of the two most promising hydrometallurgical techniques are electrokinetics and electroleaching / methods of leaching.

- *In situ recuperation. In situ* techniques have not been studied as much as *ex situ* techniques due to surface heterogeneity and longer treatment times. However, *in situ* techniques may be promising due to better cost-effectiveness and practicality.

- *Vapor extraction coupled with evaporation.* Vacuums are used in unsaturated zone to remove volatile and semi-volatile contaminants.

- *Permeable reactive barriers.* Dissolved compounds react with compounds found on the walls and then precipitate out. This technique has been employed for the treatment of sites contaminated with organic compounds and metals. These barriers are placed perpendicularly to the flow of contaminants.

- *In situ leaching and extraction.* Used together with pumps in treatment, this method uses chemicals injections to improve the solubility of mercury in groundwater.

- *Electrokinetic separation.* This process involves the generation of an electric field by applying a potential difference or current into a soil matrix. Metals such as mercury migrate towards electrodes placed in the soil where they accumulate and can be removed at a lower cost by excavating the affected area.

- *Interceptor systems.* Interceptor systems such as ditches and drains are simple and effective for the recovery of mercury as free product, but these treatments are limited by site topography and stratigraphy.
- *Phytoremediation.* Some plants have the ability to assimilate and concentrate metals in soil. The recovery of these metals occurs after collecting and incinerating the plants.
- *Passive remediation of the wetlands.* Using wetlands to immobilize mercury is a controversial topic as some wetlands contain microorganisms that can convert mercury into even more toxic species.

Containment

In the containment treatment the soils are treated on one side of the contaminated site, where the processing is performed on an area adjacent to the contaminated site or an area within the contaminated site upon removal of soil or soil-like materials. In this classification are (Hinton and Veiga, 2001):

- *Pump and treat*: With certain contaminants or systems, pollution removal is not possible and it is necessary to protect hydraulic content. When the contaminant mass remains in the subsurface, pump and treat systems can prevent site contamination.
- *Impermeable barriers (sealed surfaces and drainage)*: Mud barriers are slightly permeable barriers made of bentonite or cement-bentonite mixtures. Generally, these barriers are between 0.5 and 2 m thick and have a maximum depth of 50 m. There are other types of barriers that are constructed by injection molding or by vibratory forces. On the other hand, surface seals and drainage are used to controlling filtration and limit pollutant movement towards groundwater.
- *Stabilization and solidification*: Stabilization and solidification techniques use both *in situ* or *ex situ* conditions by mixing impacted sites. Stabilization attaches contaminants to the soil structure, which usually decreases soil permeability. Moreover, solidification improves the physical characteristics of materials such as mudor sediments; they can be excavated and transported more easily.

- *Sediment covering:* In situ covering involves placing an insulating layer over the contaminated material.

In Situ Chemical Treatment

Another option is the *in situ* chemical treatment option, which is the name of all treatments that involve the injection of a chemical reagent into an aquifer source upstream of the contaminated site. This chemical agent reacts with the contaminant, transforming it into an innocuous form; eventually, it can pump through a given volume of water which can later be recycled for injection. The following actions must be considered:

- Increase the output rate of the ground water through the contaminated zone by increasing the hydraulic gradient through injection and extraction.
- Transform the contaminant using chemical reaction within the aquifer.

Electroremediation of Polluted Soil with Mercury

Electroremediation has been successfully applied in a variety of soil restoration studies, this methodology having the advantage of exhibiting simultaneous chemical, hydraulic and electrical gradients. Indeed, for efficient mercury removal from a saturated soil with electroremediation, application of either an electric field or direct current through two electrodes (anode and cathode) is required. These are usually inserted in wells containing a supporting electrolyte made from inert salts, leading to improved electric field conductive properties (Rajeshwar et al, 1994; Huang et al, 2001; Acar and Alshawabkeh, 1993).

Furthermore, since electroremediation is a physicochemical technique based on ion transport, it is an excellent tool for the removal of inorganic species, such as Hg^{+2} (Rajeshwar et al, 1994; Bustos, 2013). The main advantages of electroremediation, as compared with other soil treatment procedures, are (Huang et al, 2001; Acar and Alshawabkeh, 1993; Ibañez et al, 1998; Segall and Bruell, 1992; Cabrera – Guzmán et al, 1990): (1) electroosmotic flow is not dependent on either pore or particle size, (2) hydraulic gradient is enhanced by electromigration, (3) treatment can be applied *in situ*, (4) it can be applied to low permeability soils, (5) there is minimal disruption of normal activities at the site, (6) the required investment is usually lower than that for other conventional treatments, and (7) it can be applied in conjunction with techniques such as pumping, vacuum extraction or bioremediation.

The processes taking place during electroremediation can be classified into two main categories: (a) processes occurring as a consequence of the applied electric potential. These processes include electromigration (ion transport), electroosmosis (mass transport), and electrophoresis (charged particle transport); (b) processes occurring in the absence of an electric potential. This includes concentration induced processes like diffusion, sorption, complexation, precipitation and acid - base reactions (Reed et al, 1995; Bustos, 2013).

Specifically, for mercury polluted soil electroremediation, the use of complexing agents like ethylendiaminetetraacetic acid (EDTA), KI, and NaCl under a constant potential gradient has been reported (Reddy et al, 2003). Based on the above precedents, the electroremediation was developed aided by extracting agents for mercury removal from San Joaquin's Sierra Gorda soil samples (Figure 5, Robles et al, 2012).

Electroremediation of mercury polluted soil, facilitated by the use of complexing agents, proved to be an attractive alternative treatment for the removal of mercury from polluted soil in mining areas located at Sierra Gorda in Queretaro, Mexico (Figure 5A and 5B). Implementation of this remediation protocol is expected to improve the living conditions and general health of the population in the Mine "El Rincón" in San Joaquin (Figure 5C). Experimental observations suggest that it is possible to remove up to 75 % of metal contaminants in mercury polluted soil samples by wetting them with 0.1 M EDTA, placing them in an experimental cell equipped with Ti electrodes, and then applying a 5 V electric field for 6 hours (Figure 5D, Robles et al, 2012). When we followed the electrochemical removal of mercury in a batch reactor (Figure 6A), it was removed around 87 % of Hg^{2+} in a time of 9 hours close to the anode side by the presence of EDTA (Figure 6B). The pH remains nearly constant at 4 and conductivity showed values close to 10 mS cm^{-1} by the ionic species.

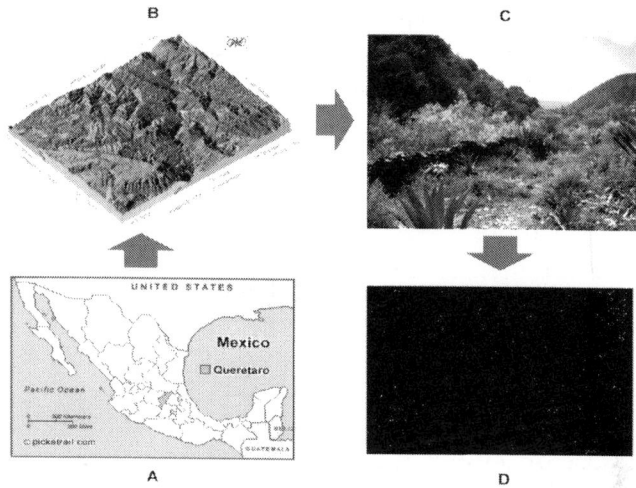

Figure 5: Localization of Queretaro in Mexico (A) with satellite image from San Joaquin's Sierra Gorda, Queretaro (B) where there is the Mine "El Rincón" (C) with high concentration of Hg^{2+}, which was removed with electroremediation process in continues flow in presence of EDTA (D).

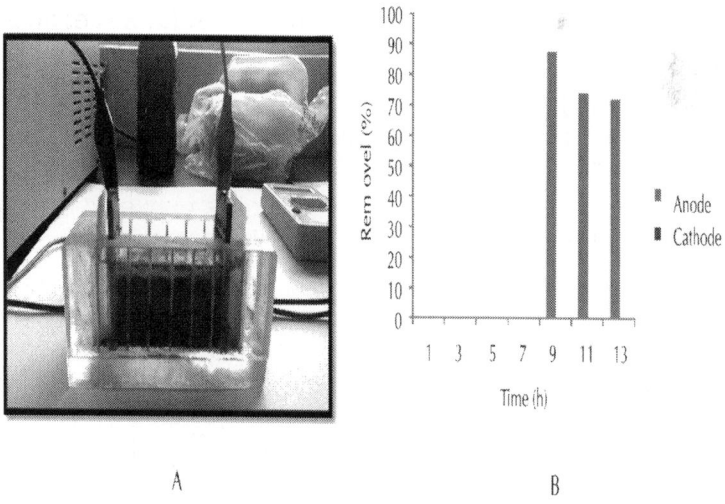

Figure 6: Electroremediation process in batch reactor assisted by EDTA (A), and its corresponding removal percentage of Hg^{2+} followed during 13 h of treatment, close to anode and cathode.

The efficient removal of mercury contaminants observed under these conditions is attributed to electromigration of the coordination complexes that form between the terminal hydroxyl groups in EDTA and divalent mercury (Hg^{+2}), which is probably strengthened by supramolecular interactions between unshared electrons at EDTA's tertiary amino nitrogens and Hg^{+2}. These interactions are particularly effective with the presence of potassium ions. This observation is supported by molecular modeling of several possible interactions in the proposed complex using the Density Functional Theory method (B3LYP LANL2DZ, Robles et al, 2012, Figure 7).

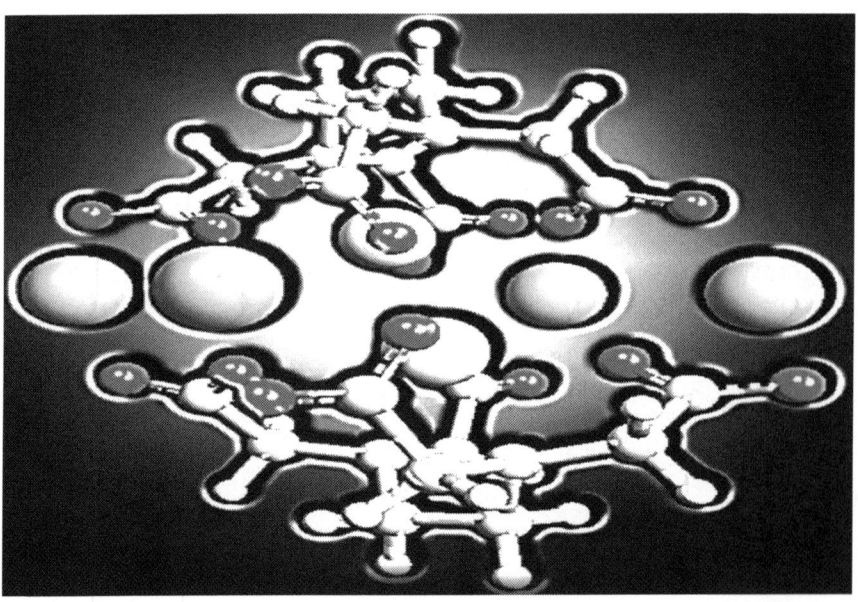

Figure 7: Optimized conformation and molecular structure of the proposed 2 Hg^{+2} / 2 EDTA / 4 Na^+ complexes (B3LYP LANL2DZ, Robles et al, 2012).

CONCLUSIONS

Mercury is a non-essential metal that can bioaccumulated in living organisms, causing toxic effects of various kinds. Therefore, it is vital to understand how this metal is transmitted through the environment and the changes that occur due to contact with living organisms, or environmental conditions such as high temperatures or strong winds. A study of the reactions that form organic or inorganic compounds, which are even more toxic, is also necessary to limit mercury toxicity. This research gives a brief overview of the techniques commonly used for treatment of sediment and soil contaminated with mercury, mobilization and loadings the mobilized mercury across different matrixes of environment, in specially electroremediation of mercury polluted soil, facilitated by the use of complexing agents as EDTA, proved to be an attractive alternative treatment for the removal of mercury from polluted soil in mining areas.

ACKNOWLEDGEMENTS

The authors would like to thank to Consejo Nacional de Ciencia y Tecnología de los Estados Unidos Mexicanos (CONACyT), L'Oreal, Academia Mexicana de Ciencias (AMC), Fundación México – Estados Unidos para la Ciencia (FUMEC) and the International Cooperation Program across Bilateral Cooperation Mexico – Hungary for the funding of this research. The contribution of J. L to this research was (partially) carried out in the framework of the Center of Excellence of Sustainable Resource Management / Applied Materials Science and Nano-Technology / Mechatronics and Logistics / Innovative Engineering Design and Technologies at the University of Miskolc. I. Robles is grateful to CONACyT for her scholarship.

REFERENCES

1.   Acar, Y. B., Alshawabkeh, A. N. (1993). Principles of Electrokinetic Remediation. *Environmental Science and Technoogy,27* (13), 2638 - 2647.

2.   Allard, B., Arsenie I. (1991). Abiortic reduction of mercury by humic substances in aquatic system- an important process for mercury cycle. *Water Soil Air Pollution,* 56, 457-464.

3.   Aposhian H. V., Maiorino R. M., Ramirez D. G., Charles M. Z., Xu Z., Hurlbut K. M., Munoz J.P., Dart. R. C. ,Aposhian M. M. (1995). Mobilisation of heavy metals by newer, therapeutically useful chelating agents. *Toxicology,* 97, 23-38.

4. Barraclough, F. R., Givelet, N.,Cortizas, A. M, Goodsite, M. E., Biester, H. Shotyk, W. (2002).An analytical protocol for determination of total mercury concentrations in solid peat samples. *The Science of Total Energy*, 292, 129-139.

5. Bengtsson, G. (2008). Mercury sorption to sediments: dependence on grain size, dissolved organic carbon, and suspended bacteria. *Chemosphere, 73*, 526–531.

6. Bernaus A., Gaona X., Ree D., Valiente M. (2006) Determination of mercury in polluted soils surrounding a chlor-alkali plant. Direct speciation X-ray absorption spectroscopy techniques and preliminary geochemical characterisation of the area. *Analytica Chimica Acta, 565*, 73-80.

7. Bollen, A. (2008). Mercury speciation analyses in HgCl2-contaminated soils and groundwater-implications for rosk assessment and remediation strategies. *Water Research, 42*, 91–100.

8. Buchet J. P., Lauwerys R. R. (1989). Influence of 2,3 dimercaptopropane-1-sulfonate and dimercaptosuccinic acid on the mobilisation of mercury from tissues of rats pretreated with mercuric chlorid, phenylmercury acetate or mercury vapours. *Toxicology*, 54, 323-333.

9. Bustos, E. (2013). Remediación Electro – Cinética de Suelos Contaminados con Hidrocarburo. Chapter 5: Remediación de Suelos y Acuíferos Contaminados en México: Bases Teóricas y Experiencias Reales. Luis G. Torres Bustillos y Erick R. Bandala González. FUNDAP / CONCyTEQ / GTZ, México.

10. Cabrera - Guzmán, D., Swartzbaugh, J. T., Weisman, A. W. (1990). The Use of Microscale in Hazardous Waste Site Remediation. *Journal of Air and Waste Management Association*, 40, 1670 – 1676.

11. Caldwell C. A., Canavan C. M., Blom N. S. (2000).Potential effect of forest fire and storm flow on total mercury and methyl mercury in sediments of an arid-lands reservoir. *The Science of the Total Environmental*, 260, 125-133.

12. Evangelu M.W.H., Ebel, M., Schaefer, A. (2007). Chelate assisted phytoextraction of heavy metals from soil. Effect, mechanism, toxicity and fate of chelating agents. *Chemosphere*, 68, 889-1003.

13. Fergusson, J. E. (1990). The Heavy Elements: Chemistry, Environmental Inpact and Health Effects, Pergamon Press

14. Fügedi, U, J. Vatai and L. Kuti (2011). Mercury Content in the Superficial Geological Formations of Hungary, Central European Geology Vol. 52 (3-4), 287-298

15. Guzzi G., Minoia C., Pigatto P.: Fatal mercury poisioning and chelating agents. Letter to the Editor. Forensic Science Intern. 202, (2010), 61.

16. Han, F. X., Su, Y., Monts, D. L., Waggoner, C.A., Plodinec, M.J. (2006).Binding, distribution and plant uptake of mercury in a soil Oak Ridge, Tenessee, USA.*Science of the Total Environment*, 368, 753-768.

17. Harvey, J. W. (2002). Interactions between surface water and ground water and effects on mercury transport in the north-central everglades. Reston Virginia: USGS, Science for changing world.

18. Hinton, J.,Veiga, M. (2001). Mercury contaminated sites: a review of remedial solutions. National Institute for Minamata Disease. In: Proceedings of the NIMD (National. Institute for Minamata Disease)

19. Holley, E.A, Mc. Quillan, A. J., Craw, D., Kim, J. P, Sander S. G. (2007).Mercury mobilization by oxidative dissolution of cinnabar and metacinnabar. *Chemical Geology*, 240, 313-325.

20. Hylander, L. D.,Meili, M. (2003). 500 years of mercury production: global annual inventory by region until 2000 and associated emissions. *The Science of Total Environment*, 304,13-27.

21. Huang, C. P., Cha, D., Chang, J. –H, Qiang, Z. (2001). Electrochemical Process for in-situ Treatment of Contaminated Soils, Newark, Delaware.

22. Ibanez, J. G.; Singh, M. M.; Szafran, Z.; Pike, R. M. (1998). Laboratory Experiments on Electrochemical Remediation of the Environment.Part 3.Microscale Electrokinetic Processing of Soils. *Journal of Chemical Education*, 75 (5), 634 - 635.

23. Iverfeld, A., Lindquist, O. (1986).Atmospheric oxidation of elemental mercury by ozone in the aqueous phase.*Atmospheric Energv.* 20, 1567-1573.

24. Ko, K. B., Byun, Y., Cho, M., Namkung, W., Shim, D. N.., Koh, D. J., Kim, K. T. (2008). Influence of HCl on oxidation of gaseous elemental mercury by dielectric barrier discharge process. *Chemosphere, 71*, 1674-1682.

25. Lakatos, J., Brown, S. D.,Snape, C. E. (1999). Influence of coal properties on mercury uptake from aqueous solution. *Energy and Fuel, 13*, 1046-1050.

26. Lakatos, J., Akcin, G., Brown, S. D., Snape, C.E. (1999). Application of coal and biomass type sorbents for Hg(II) and Cr(VI) removal in the environmental protection technology. Challenges of an inderdisciplinary Science, AcademiaiKiado, Budapest, 327-336.

27. Lakatos, J., Cheng-gong, S. C., Perry, R., Kennedy, M., Snape, C. E. (2009). Ultra high capacity co-precipitated manganese oxide sorbents for oxidative mercury capture. 237 th ACS meeting Spring, Salt Lake City.

28. Leopold, K., Foulkes, M., Worsfold, P. (2010). Methods for the determination and speciation of mercury in natural waters- a review. *Analytica Chimica Acta, 663* (2), 127–138.

29. Lomonte, C., Doronila, A.,Gregory, A., Baker, A.J.M., Kolev, S.D. (2011). Chelate –assisted phytoextraction of mercury in biosolids. *Science of Total Environment, 409*, 2685-2692.

30. Lopez –Anton, M. A., Yuan, Y., Perry, R., Morato –Valer, M. (2010). Analysis of mercury species present during coal combustion by thermal desorption. *Fuel 89*, 629-634.

31. Lopez –Anton, M., Perry, R., Abad – Valle, P.,Diaz – Somano, M., Martinez –Tarazona, M. R., Morato-Valer, M. (2011).Speciation of mercury in fly ashes by temperature programmed desorption. *Fuel Processing Technology, 92*, 707-711.

32. Moreno, F. N., Anderson, C.W.N., Stewart, R.B., Robinson, B.H. (2004). Phytoremediation of mercury contaminated mine tailings by induced plant-mercury accumulation. *Environmental Practice, 6*, 165-175.

33. Nik, M. G. (2012). The estudy of mercury pollution distribution aroun a chlor-alkali petrochemical complex, Bandar Iman, southern Iran.*Environmental Earth Science, 67* (5), 1485 – 1492.

34. Nriahu J. O. (1994).Mercury pollution from the past mining of gold and silver in Americas. *The Science of Total Environment*, 149, 167-181.

35. Orecchio, S., Polizzotto (2013). Fractionation of mercury in sediments during draining of Augusta (Italy) coastal area by modified Tessier method.*Microchemical Journal*, 110, 452-457.

36. Okouchi, S., Sasaki, S. (1984). Volatility in mercury in water. *Journal of Hazardous Materials*, 8, 341-348.

37. Ozerova, N. A. (1996). Mercury in Geological Systems In: W. Baeyens, R. Ebinghaus and O. Vasiliev (eds) (1996). Global and Regiopnal Mercury Cycles: Sources, Fluxes and Mass Balances, NATO ASI Series 2. Environment Vol. 21 Kluwer Academic Publishers Dondrecht/Boston/London, 463-474

38. Pongratz, R., Heumann, K. G. (1988).Production of methylated mercury and lead by polar macroalgae. A significant natural source for atmospheric heavy metals in clean room compartments. *Chemosphere*, 36, 1935-1946.

39. Pillay, K., Cukrowska, E. M., Coville N.J. (2013).Improved uptake of mercury by sulphur-containing carbon nanotubes. *Microchemical Journal*, 108, 124-130.

40. Rajeshwar, K., Ibanez, J. G., Swain, G. M. (1994). Electrochemistry and the environment, *Journal ofApplied Electrochemistry*, 24,1077–1091.

41. Reed, B. E., Berg, M. T., Thompson, J. C., Hatfield, J. H. (1995). Chemical conditioning of electrode reservoirs during electrokinetic (EK) soil flushing of a Pb- contaminated silt loam, *Journal of Environmental Engineering* ASCE, 121 (11)805 - 815.

42. Reddy, K. R., Chaparro, C., Saichek, R. E. (2003). Removal of Mercury from Clayey Soils Using Electrokinetics, *Journal of Environmental Science and* Health, Part A – Toxic / Hazardous Substances & Environmental Engineering, A38 (2), 307 - 338.

43. Reddy, K. R., Chaparro, C., Saichek, R. E. (2003). Iodide – Enhanced Electrokinetic Remediation of Mercury – Contaminated Soils, *Journal of Environmental Engineering,* ASCE, 129 (12),1137 – 1148.

44. Robles, I., García, M. G., Solís, S., Hernández, G., Bandala, Y., Juaristi, E. and Bustos E. (2012). Electroremediation of mercury

polluted soil facilitated by complexing agents, *International Journal of Electrochemistry Science, 7*, 2276 – 2287.

45. Robles – Gutiérrez, I., Solís – Valdéz, S., Hernández – Silva, G., Bustos, E. (2012). Electroremediation of Mercury Polluted Soil by Complexing Agents. Environmental Influences of Mercury Ore Processing: Case Studies Selected at Slovenian, Mexican, Hungarian Group Meeting in Idrija, Editors: MatejaGosar, Tatjana Dizdarevič, Miloš Miler, UNESCO, Slovenian National Committee of the International Geoscience Programme, Idrija, Slovenia.

46. Segall, B. A.; Bruell, C. J. (1992). Electroosmotic Contaminant Removal Processes, *Journal of Environmental Engineering*, 118 (I), 84 - 100.

47. Scharek, P. Hernández–Silva, G. Solorio–Munguia, G. Vassallo–Morales, L. Bartha, A. Soliz–Valdez, S. Tullner, T. (2012). Total Mercury Content In Soils, Sediments and Tailings in San Joaquin, Querétaro, Mexico Annual Report of the Geological Institute of Hungary, 2010, 125–129

48. Shi, J. B., Liang, L. N., Jiang, G. B., Jin, X. L. (2005). The speciation and bioavailability of mercury in sediments of Haihe River, China. *Environment International, 31*, 357– 365.

49. Shon Z. H., Kim K.H., Kim M.Y, Lee M. (2005). Modelling study of reactive gaseous mercury in urban air.*Atmospheric Environment*, 39, 749-761.

50. Slowey, A. J., Rytuba, J. J., Brown, J. G. E. (2005). Speciation of mercury and mode of transprot from placer gold mine tailings. *Enviromental Science and Technology*, 39 (6), 1547-1554.

51. Somoano M. D., Unterberger S., Hein K. R. G. (2007). Mercury emission control in coal-fired plants: The role of wet scrubbers. *Fuel Processing Technology*, 88, 259-263.

52. Sondreal, E.A., Benson, S.A., Pavlish, J.H., Ralston N.V.C. (2004). An overview of air quality III. Mercury, trace element and particulate matter.*Fuel Processing Technology*, 85, 425-440.

53. Wang M., Keener T. C., Khang S.J. (2000).The effect of coal volatility on mercury removal from bituminous coal during mild pyrolysis. *Fuel Process Technology*, 67,147-161.

54. Wartel, M., Mikac, N., Ouddane, B., Niessen, S. (1999). Speciation of mercury in sediments of the Seine estuary (France). *Applied Organometallic Chemistry,* 13 (10), 715 – 725.

55. Weber J. H. (1993). Review of possible path for abiotic methylation of mercury(II) in the aquatic environment. *Chemosphere,* 26, 2063-2077.

Optical Fibers to Detect Heavy Metals in Environment: Generalities and Case Studies

J. A. García[1], D. Monzón[2], A. Martínez[2], S. Pamukcu[3], R. García[4], and E. Bustos[1]

[1]Centro de Investigación y Desarrollo Tecnológico en Electroquímica S.C., Querétaro, México

[2]Centro de Investigaciones en Óptica A. C., León, México

[3]Fritz Engineering Laboratory, Lehigh University, Bethlehem, USA

[4]Laboratorio de Química Atmosférica, Centro de Ciencias de la Atmósfera, Universidad Nacional Autónoma de México, Ciudad Universitaria, Coyoacán, Mexico

INTRODUCTION

The fiber optic sensors can be used to create a truly distributed chemical sensing capability for selectively detecting metal compounds by spatial and temporal acquisition over large distances in the subsurface. In addition the fiber optic sensors have several advantages such as small size, light weight, immunity to electromagnetic interference (EMI), high temperature performance, large bandwidth, high sensitivity, and environmental ruggedness (Krohn, 1988). Most current technologies capable of detecting contaminants use strategically placed sensing or monitoring devices. This works reasonably well if plausible event location is known, hence settle recording vast amounts of benign data over time until the appearance of the suspected event. This approach remains limited for application in large spatial scales in the geo-environment and subsurface. A simple approach is to suppress all the benign data by triggered transmittal of the signals only at the spatial and temporal vicinity of the event. This, in essence the "truly distributed" sensing capable of delivering the event signal "*wherever*" and "*whenever*" it might occur, as opposed to only at strategic places where the sensing devices are pre-located. The revolutionary advances in flexible sensing and distributed data processing permits us sensing in this truly distributed manner.

Sensors based on fiber optic cable functions make use of the following important features of the cable to sense the environment: (1) optical loss: intrinsic and extrinsic energy loss properties, (2) refractive index: index profile in radial direction and the reduction of index fluctuation along the axial direction; (3) shape: cross sectional shape and size, the surface finish and the fluctuation of the size along axial direction. Present fiber optic sensors mostly use energy loss principles (i.e., changes in optical power in linearly positioned wave-guides) for chemical detection. These can be limited for distributed applications if energy depletes over a short stretch of the fiber sensor, or frequent sensor points are needed at a prohibitively expensive cost. Other sensors use the changes in refractive index and/or cross sectional size of the fiber cable that change the light scattering property in optical fibers, known as Brillouin scattering (Horiguchi et al, 1995; Kee et al, 2000).Fiber optic sensing based on Brillouin scattering has been used successfully in civil infrastructure for health monitoring (Bao et al, 2001; Ohno et al, 2001). In this chapter a background on use of optical fibers for chemical sensing and new developments and proposed advancements are discussed.

BACKGROUND

Sensors Overview of Fiber Optic

In an optical fiber sensor a physical, chemical or biological variable can interact with the light and produce a change in one of their parameters. It is desirable to produce an optical signal related uniquely to the parameter of interest. These sensors use the optical fiber either as the sensing element (intrinsic sensors), or as a means of relaying signals from remote sensing area to the signal processor (extrinsic sensor), or both. Optical fiber sensors take advantage of the inherent fiber optic characteristics which include their lightweight, of very small size, passivity, low-power requirement, resistance to electromagnetic interference, environmental impact and corrosion, their bandwidth, and flexibility. They can be installed in areas normally inaccessible by conventional sensors, they can be interfaced with data communication systems and pose no risk of electric shock in live measurements. These attributes have allowed optical fiber sensors to displace traditional sensors for measurement and monitoring of rotation, acceleration, electric and magnetic field, temperature, pressure, acoustics, vibration, linear and angular position, strain, humidity, viscosity, pH, gas and chemical content among many others.

Use of Fiber Optic Sensors is a viable real-time data gathering approach by surface-adhering or embedding the fiber to a specimen under evaluation. The concept of embedding fiber-optic sensors into structures has generated a great deal of interests in aerospace engineering initially and more recently in civil engineering. There are several types of chemical sensing techniques based on optical waveguides (Ho et al, 2001). Among those are fiber Bragg gratings (FBG), which is marking of a fiber with a laser to create a local narrow band pass filter sensitive to environmental parameters (Guemes et al, 1998; Schulz et al., 1998). Optical time domain reflectometry (OTDR) consists of sending a powerful light pulse and observe modification in the reflected light due to local in homogeneties along the fiber. The pulse losses correspond to specific environmental interaction. The evanescent pulse technique is also based on OTDR, in which the fiber cladding is modified to interact with the environment and the pulse travels partially through the cladding. These sensors demand large optical power, due to the cumulative energy loss at the points of contact with the chemicals.

Over the last decade, there has been rapid development in the area of smart sensor technologies, in particular using structurally integrated optical fiber to form the basis for smart structure technology. A variety of configurations have been developed for measurement of strains and deformations in structures, including localized-type such as fiber Bragg gratings and multiplexed long gauge interferometric sensors, and distributed sensing schemes including Stimulated Brillouin Scattering (SBS) or Brillouin Optical Time Domain Analysis (BOTDA) (Bao et al, 2001) and Brillouin Optical Time Domain Reflectometry (BOTDR) (Pamukcu et al, 2006; Anastasio et al, 2007).

Between different types of optical sensors reported, there are those based on sensitive coatings onto the fiber surface, Fabry-Perot interferometers, long-period fiber gratings (LPFG), LPFG with sensitive films, hetero-core devices, fiber Bragg gratings on doped fibers (i.e, Germanium doped). Fiber gratings are structures consisting of a periodic perturbation of the optical and/or geometrical properties of an optical fiber. Depending on the pitch of the perturbation, fiber gratings fall into two distinct categories: short period gratings, known as fiber Bragg gratings (FBGs) and, long period gratings (LPFGs). Stretching the fiber gratings causes a change in grating period, hence the wavelength of the reflected light. This makes the FBGs ideal for localized temperature and strain measurements. Unlike FBGs in which counter directional coupling occurs in the core, co-directional coupling occur in LPFGs between the core and cladding. This feature renders LPFGs sensitive not only to temperature and strain, but also to bending causing a curvature, to hydrostatic pressure, to torsion and to ambient refractive index changes. The closer the ambient refractive index to that of the cladding the stronger the sensitivity to refractive index changes. It is this high sensitivity that has piqued the interest in development of various types of refractive index-based LPFG sensors which constitute most of the chemical sensing applications (Orellana and Haigh, 2008; Kasik et al, 2010).

Point detection fiber optic sensors have been developed successfully for measurement of liquid levels, chemical species, drugs, environmental agents (such as pollutants and pesticides), biochemical reactions, and to monitor a wide variety of chemical processes (Wolfbeis, 2000). A fiber optic laser induced breakdown spectroscopy method was demonstrated in the field using a push-cone device, which is a single point, single time measurement technique. The most common configuration for optical pH sensors, and other environmental parameters, employs a fluorescence indicator (Lee et al, 2000). Among the different types of optical fiber devices used in pH sensing are, hetero-core fibers, U-bend fibers, fiber Bragg and long-period gratings, fibers and fiber tips with active doped cladding, among others (Kocincovaet al, 2007). Some of the substances that can be detected or identified using optical fiber sensors are volatile organic compounds (alcohols, formaldehydes, methane, ketones, CO_x, O_2, and H_2), some metallic ions like Ca, Al, Cu, Zn, Hg, V and Pb (Jeronimoet al, 2007; Wolfbeis, 2008).

Wide application of advanced chemical sensing in the environment may suffer from scaling issues. The real-world conditions often require self-referencing, spatially distributed, temporally continuous, and chemically selective sensors for monitoring regions spanning over long lengths or wide areas. When large area monitoring for chemical agent intrusion is required, use of currently available point sensors can be cost prohibitive. Other non-point, distributed detection methods based on energy loss principles (Buerck et al, 2001) may also be inadequate when scaled to wide area monitoring due to extensive energy input requirements.

One of the unique features of the optical fiber technology is the possibility to construct distributed sensors, in which the measuring can be determined along a line of space with a given spatial resolution (Galindez-Jamioy et al, 2012) by, for example, Brillouin optical time domain analysis (BOTDA) (Cui et al., 2009, 2010 and 2011); an hetero-core LPFG sensors. In here, we examine current and proposed application of these techniques to spatially distributed, temporally continuous, and chemically selective sensing applications in soil and water environment. The premise of Brillouin technique goes back to 1920 when physicist Leon Brillouin first studied the diffusion of light by acoustic waves. The phenomenon he observed was a frequency change of scattered light. The first major papers related to distributed fiber optic sensor based on Brillouin were generated in mid-nineties (Bao, et al. 1995;Fellay et al, 1997). Current research on Brillouin sensing may be divided in three categories: photonics (the physics of

Brillouin); data processing and post processing to improve signal to noise ratios, and applications of distributed sensing to civil infrastructure and environment.

Fundamentals of Optical Techniques

The interest on optical techniques to measure or detect chemical agents have been continuously extending and growing over the last forty years. Special attention has been focused on the development of optical sensor to detect heavy metals, due to the hazardous effects of these ions on the health of human beings and ecosystems. Optical methods have the advantage of being fast, simple, compact, portable, low-cost, with sensitivities and resolutions improved to detect in the picomolar range.

Combined with other technologies, like microfluidics systems, optical waveguides, or MEMs, optical methods are suitable for application where conventional electrodes cannot be used because of their large size or because of the risk of electrode shock during in vivo measurements. Due to their minute size; these optical microsystems are capable of gathering diverse data with a small amount of analyte. The diversification of optical techniques have made possible to construct novel sensing platforms to detect heavy metals in air, water or soil, food and beverages, or biological samples.

Optical sensors to detect heavy metals employ an optical transduction technique, i. e. an element that "translate" the chemical variable into an optical signal (intensity, wavelength, polarization or phase), to yield analyte information (McDonagh et al, 2008, Grattan and Meggitt, 1999). Optical chemical sensors can be categorized, according to the transduction technique, in direct sensors and reagent-mediated sensing systems. In direct sensors the element of interest is detected directly via an optical property of the sample such as scattering or florescence, for example.

However, most heavy metals optical sensor uses an intermediate agent. Most of the optical chemical techniques to detect heavy metals are based on optical absorption, fluorescence, Raman spectroscopy, or surface plasmon resonance, whereby the perturbed signal is related to the reaction of the intermediate agent under the presence of a specific heavy metal. In general, all these techniques involves the interaction of an incident beam over an analyte or indicator element yielding transmitted, reflected or fluorescent signal. A schematic representation of the spectroscopic principle, the working mechanism of an optical sensor is shown in Figure 1.

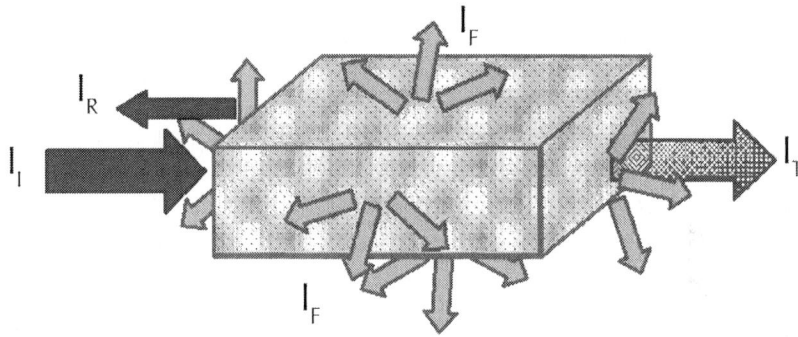

Figure 1: Representation of the optical signals in an optical chemical sensor. The incident beam I_I interacts with the heavy metal sensitive layer and depending on the optical properties of the sample one of the three signals (transmitted I_T, reflected I_R, and generated by fluorescence or scattering I_F) will be produced and will give information about the type and concentration of heavy metal present on the sample.

Absorbance-Based Techniques

Among the optical chemical techniques, the simplest to implement is that based on the measurement of light absorbed by a sensitive heavy metal layer. Absorption in a gas or liquid, where it is assumed that each single molecules equally contributes to the total light absorbed, may be characterized by a Beer-Lambert law, or simply the Beer law,

$$I_T = I_I 10^{-\varepsilon C d} \tag{1}$$

where I_T and I_I represents the intensity of the transmitted and incident beam, ε is the molar absorptivity ($L mol^{-1} cm$), and C is the concentration (mol L^{-1}) of the absorbing species and d is the absorption path length (cm). In the case of a solid, absorbing and homogeneous medium, the transmitted signal is calculated using the Lambert Bouguer law, expressed as

$I_T = I_i e^{-\alpha x}$, where x is the thickness of the medium and α is the extinction coefficient. The Beer law can also be expressed in terms of the absorbance (or optical density) A:

$$A = \log_{10}\left(\frac{I_i}{I_T}\right) = \varepsilon Cd \qquad (2)$$

There exists a linear relation between the absorbance and the concentration of the element to be measured. However, in order to observe the linear dependence of absorbance on concentration, the incident beam should be ideally monochromatic. In the case that a wide broadband light source is used, the contribution of all wavelengths must be considered, in such cases the equation (2) becomes:

$$\overline{A} = \log_{10}\left(\frac{\int I_i(\lambda)d\lambda}{\int I_T(\lambda)d\lambda}\right) \qquad (3)$$

Also the presence of highly absorbing or highly scattering media should produce a deviation from perfect Beer law behavior. In the case that more than one absorbing material is present the absorbance contribution of each species must be considered. In most of the absorbance-based heavy metals sensors an intermediate agent, an optical film that changes its absorbance according to the concentration of a specific heavy metal, is used (Antico et al, 1999; Guo et al, 2006).

A special case of the absorption-based sensors are those schemes where materials that change their color under the presence of a specific heavy metal are used (Balaji et al, 2006; Prabhakaran et al, 2007). The reaction of the sensitive components to the concentration of a specific ion produces a photochromic reaction that can be observed with a naked eye. Such materials are often in solution, but for sensing the most attractive are those that can be deposited as thin films over a substrate. The instrumentation of absorption-based sensors is the simpler of the optical heavy metal techniques, since it can be implemented with a monochromatic light source and a photodetector. This also makes this technique very susceptible to be implemented in microscopic opto-fluidic configurations that could diversify the technique.

Reflectance-Based Techniques

It is well-known that chemical reactions could lead to changes in the complex refractive index of a substance; this fact has been impulse researchers to design and fabricate materials that react with heavy metals that can be used as transducers. When these materials, commonly in the form of a thin layer, are illuminated with an appropriated light the signal will be partially or totally reflected. However, this reflectance will change when the layer is in contact with a specific metal that it reacts with. If the refractive index of the layer is purely real, the changes in the reflected signal can be estimated by using the Fresnel formulae.

However, in most cases the optical response of these materials under the presence of heavy metals are more complex and involve a change in the real and imaginary parts of the refractive index, that produce changes in reflectivity and absorbance. Also, there is a contribution of scattered light. So, the reflected signal is composed of light from different sources, however, also in this complex response the signal reflected is used to deduce, directly or indirectly, the concentration C of the heavy metals.

The reflected-based techniques are specially used in optical fiber schemes since the set-up is very simple to implement (Yusofand Ahmad, 2003, Guillemain et al, 2009). The material sensitive to the heavy metals are directly deposited over the fiber tip or in a substrate that will be illuminated by an optical fiber. The reflected signal is usually collected by the same fiber, but frequently another fiber or fibers are used to collect it. The reflected signal propagates along the fiber to the detector, where it is analyzed in order to determine the heavy metal present and their concentration (Figure 2).

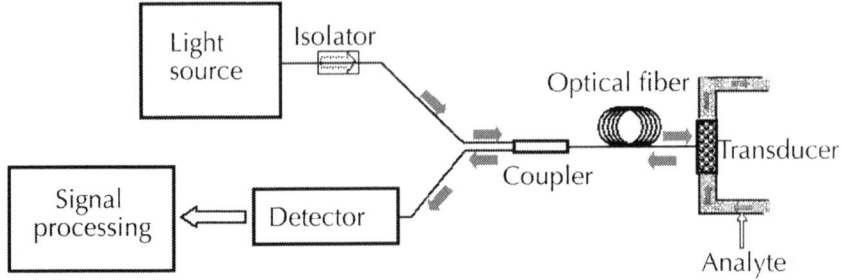

Figure 2: Diagram of a reflectance-based optical chemical sensor based on optical fibers.

Fluorescence-Based Techniques

Some materials have the property of being fluorescent when they are illuminated with a light source of appropriated wavelength. The fluorescence is the optical radiation generated when electrons of an atom or molecule return from the excited to the ground state after absorption of a photon from an excitation light source. In general the energy of the excited photon is lower than the absorbed one so the wavelength of the fluorescence signal is longer than that of the excitation.

The intensity of the fluorescent signal (I_F) is proportional to the intensity of light absorbed by the sample (I_i-I_T), therefore it is possible to establish a direct relation between the intensity of the fluorescent signal and the concentration of an absorbing material. This feature is very important for sensing since intensity of the fluorescence increases as the concentration of the absorbing species augments. Although, we have just made reference to the fluorescence intensity, for sensing, the decay time of the fluorescence signal is more frequently used because this parameter is less sensitive to source fluctuations, interference from ambient light or drift due to aging of detector. It is possible to design and fabricate a fluorescent material sensitive to a specific heavy metal. Thus, the intensity, wavelength and life time of the fluorescent signal will change under the presence of this metal. Fluorescence-based techniques are the most used to detect the presence of heavy metals due to its extraordinary sensibility (Mayra et al., 2008;Achatz et al, 2011, Aksuner, 2011).

Surface Plasmon Resonance-Based Techniques

The most popular label-free refractometric technique is the Surface Plasmon Resonance (SPR), since it allows the direct observation of chemical reactions in real time without the use of markers or labels. SPR is a quantum optical-electrical phenomenon produced by the interaction of light with a metal surface. Actually, the surface plasmon is a charge density oscillation that exists at a metal-dielectric interface. The plasmon propagates in a direction parallel to the metal-dielectric interface in the boundary of the metal and the external medium (Figure 3).

These oscillations are very sensitive to any change in the optical refractive index of the material at the boundary. The optical excitation of plasmon can be achieved in a three-layer system consisting of a thin metal film sandwiched between two isolators of different dielectric constant (Maier, 2007), where the phase-matching condition between the optical and plasmon wave vector is fulfilled. In the optical domain, the surface plasmon excitation will be observed as an intensity transmission loss at a specific wavelength. The wavelength of the dip depends on the refractive index of the two dielectrics and the thin metal film, and the propagation constant of the optical waveguide. There are three common method to excite surface plasmon, using a prism coupler and the attenuated total reflection, a periodic grating, and an optical waveguide planar (Figure 3A) or cylindrical. The prism coupler technique is the most popular since exhibits a good sensitivity, stability, and reproducibility for the measurement of heavy metals (Forzani et al, 2007; Lin et al, 2009; Abdi et al, 2011; Fen et al, 2012 and 2013; Fen and Yunus, 2013). For heavy metal detection a sensitive thin film layer is deposited over the thin metal film, so when the target heavy metal interacts with the layer a refractive index change is produced. The surface plasmon conditions changes and the peak wavelength shifts as can be seen in Figure 3B. SPR is the most sensitive refractometric method, with a theoretical resolution of 1×10^{-7}, so it is possible to detect very small traces of heavy metals.

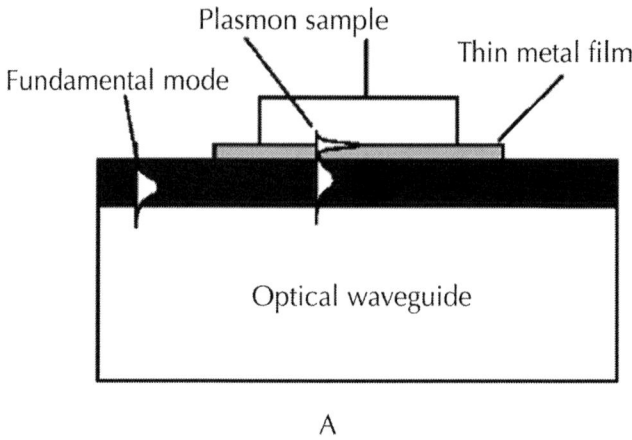

A

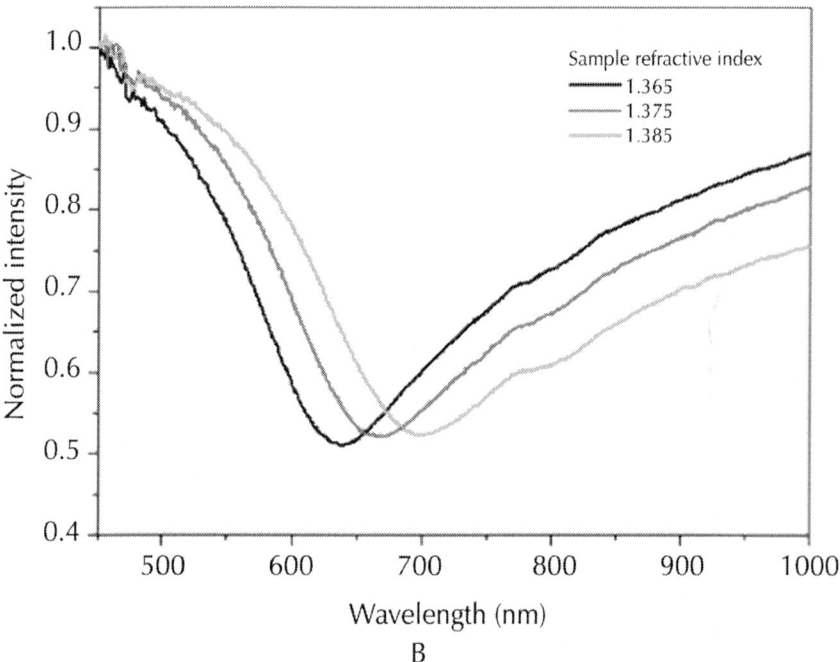

B

Figure 3: (A) A schematic representation of the surface plasmon excitation in a dielectric planar waveguide coupler configuration. (B) Optical transmission spectra of an hetero-core optical fiber coated with a 20 nm thin gold film when the fiber is immersed in a liquid with a refractive index of 1.365, 1.375, and 1.385, black, red and green line respectively.

Fundamentals of BOTDA and LPFG Based Sensing

Figure 4, 5 and 6 show the stimulated Brillouin scattering based BOTDA photonics configuration and the principle of measurement used at Lehigh University Geo-sensing laboratory, respectively (Texier et al, 2005; Pamukcu et al, 2006; Turel and Pamukcu, 2006; Anastasio et al, 2007). Brillouin is a nonlinear effect, in which light is scattered at well-defined points along the fiber where the acoustic properties of the fiber are locally modified by the environment. The stimulated Brillouin scattering (SBS) it is an acoustic – optical process which is useful for distributed measurements of a probe beam by the SBS interaction with a counter-propagating nanosecond pump pulse. In the SBS technique, as in a null detector, the pump and probe are initially de-tuned by a (frequency) that is slightly greater than the Brillouin frequency. Therefore, in unstressed fiber, the base line remains flat resulting in a self-referenced sensor eliminating the need for duplication with another reference fiber.

The Brillouin line being intrinsically narrow (~20 - 50 MHz), the initial de-tuning can be quite small so that the amount of strain required to generate a signal is also quite small (0.001 %), allowing for higher resolution and sensitivity of the sensor compared to other fiber-based measurement techniques. When SBS based sensors are used for environmental sensing the fiber is hitched or bonded with selective polymer transducers that are mass detectors in direct contact with the surrounding medium. The polymer reacts to the surrounding (i.e. moisture, pH, target chemical) by selectively absorbing the target compound and it swells. Localized swelling of the bonded or hitched polymer produces tangential, axial or radial stresses on the fiber depending on the physical coupling. These stresses result in axial straining of the fiber and a measurable change in its local acoustic properties, hence a Brillouin scatter of the transmitted light. The location of the generated signal is determined by time domain reflectometry.

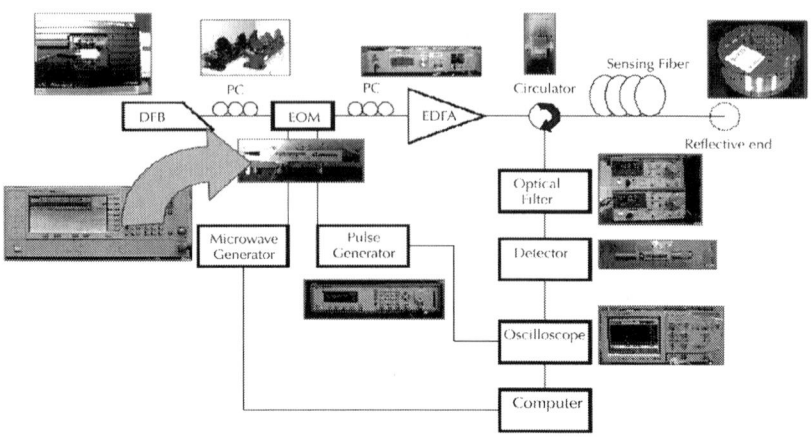

Figure 4: SBS based BOTDR photonics set up at Lehigh Geo-Sensing Laboratory.

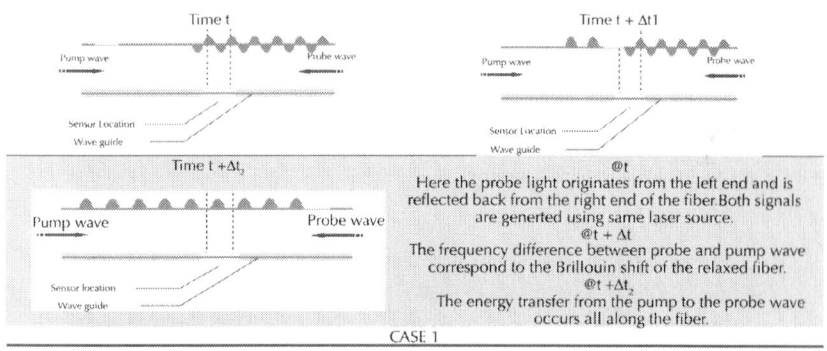

Figure 5: Measurement principles of SBS based sensing: case I.

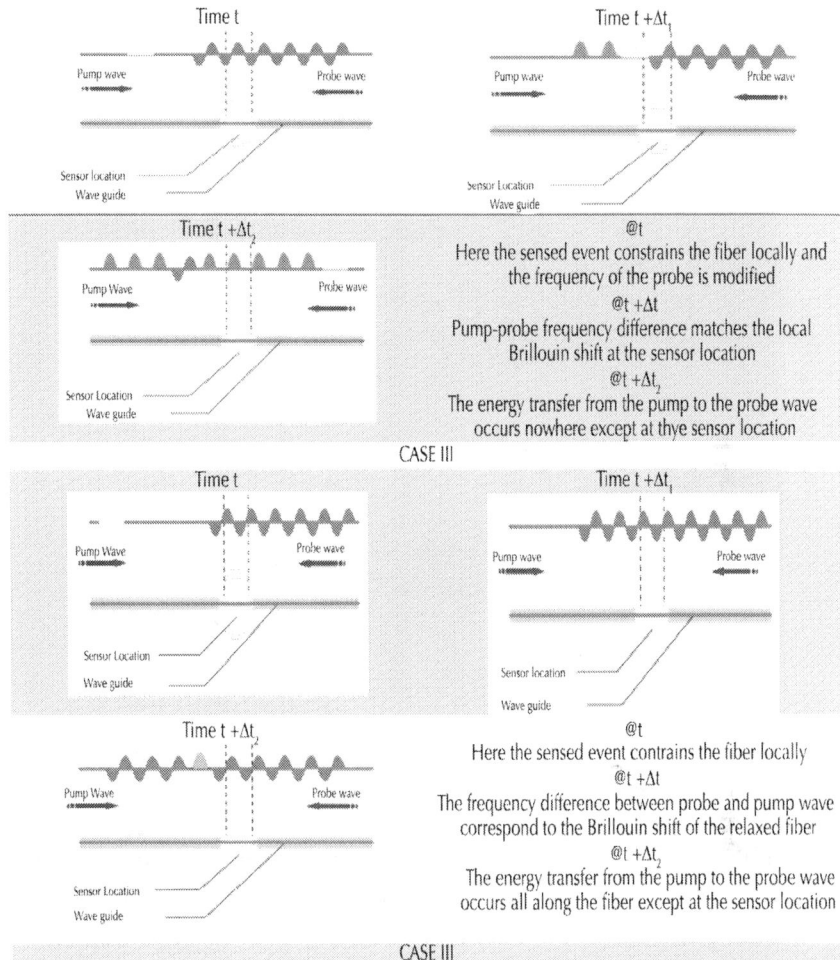

Figure 6: Measurement principles of SBS based sensing: case II and III.

In the case of LPFG`s the environmental changes produce a spectral shifts, that required a sophisticate or a complicate equipment. These devices are very sensitive to temperature changes so to measure another different parameter it is necessary to make the corresponding compensation. One alternative to avoid these difficulties are hetero-core fibers. These devices are constructed by changing the diameter of the core in a small length (mm) in a transmission line (Figure 5 and 6), which causes the optical wave to expand within the cladding in the single-mode region of the hetero-core, thus the evanescent field can easily interact with the external medium. Owing to the core diameter mismatch, some of the light is guided by the cladding of the SM fiber (Figure 7). This makes the transmission of the device dependent on the refractive index of the external medium. The sensor exhibits maximum transmission changes when the index of the sample medium approaches that of the SM fiber cladding. The device can operate at different wavelengths as well as when coated with thick films made of variable index materials. Moreover, standard emitters, fibers, detectors, etc., are needed to fabricate the sensor, which makes it attractive for diverse applications (Villatoro and Monzon-Hernandez, 2006).

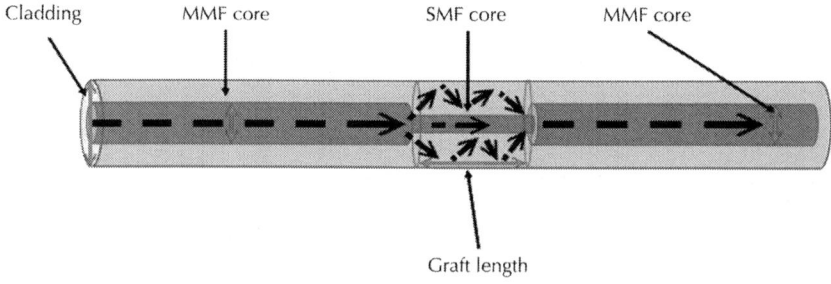

Figure 7: Schematic representation of the hetero-core fiber structure. A piece of single-mode fiber (SMF) of length L is inserted between two multimode fibers (MMF). Cladding diameter of both fibers is 125 μm but the core diameters are different.

INVESTIGATION

BOTDA/BOTDR Applications and Proposed Sensor Configurations

The usefulness of SBS for sensing is general: any change in external conditions, that affects the acoustic properties of the optical fiber, can in principle be detected. This is true of direct temperature and pressure changes, but can also be true of changes in chemical environment that can be made to result in temperature and pressure changes. An

SBS based BOTDR sensing system was used to detect water content changes in soil. Water transducers (hydrophilic polymers) were tested to correlate Brillouin strain response to the water content of the surrounding soil environment. In these experiments, the optical fiber was wound and secured about discretely placed discs (2 cm length x 5 cm diameter) of AEP60 hydrophilic polymer (Figure 8A), stringed along 100-m fiber continuous optical fiber. The diameter of the polymer disc was selected to accommodate the minimum curvature of bending of the fiber, as shown in Figure 8B.

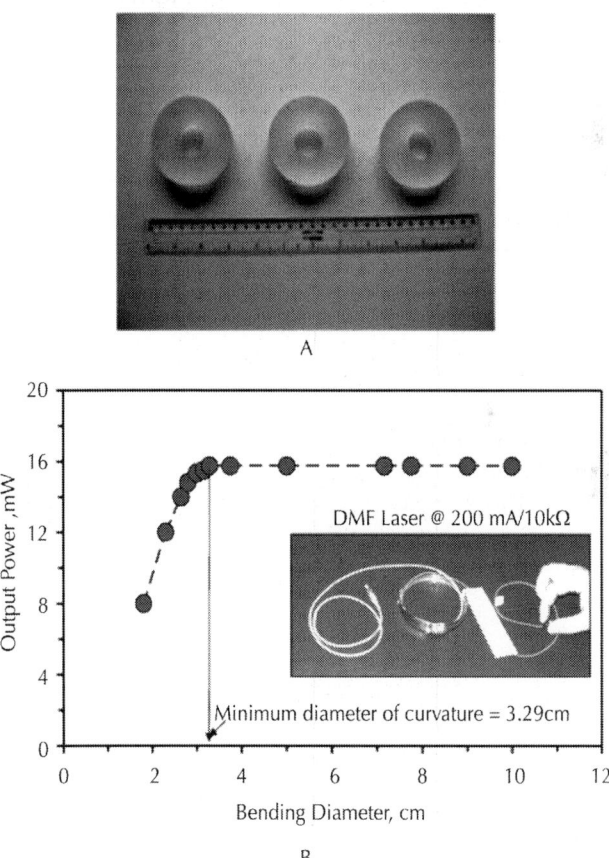

Figure 8: A photograph of the AEP60 hydrogel polymer discs used in sensor (A); assembly and test results for minimum fiber bending radius assessment (B).

The optical fibre at the inlet and outlet of the string of four transducers were spliced to spools of fibre on each end, and connected to the photonic set-up. Each water transducer was then embedded in a wet clay sample of predetermined water content (5, 10, 20 and 30 % by dry weight of clay), as shown in the inset sketch of Figure 9B.

The clay samples were packed in equal volume, watertight, cylindrical cells of 14-cm diameter and 28-cm height. The experiments were conducted in a temperature-controlled environment, at 25°C so that Brillouin scattering measurements were not influenced by thermal expansion or contraction of the fibre. The Brillouin shift was measured with 5 to 20 minute intervals up to the maximum observable swelling. When no significant change in Brillouin shift was recorded for three consecutive measurements, the transducers were removed from the soil chambers and left for open air-drying.

The Brillouin shift measurements were recorded with 5 to 20 minute intervals until the drying phase was completed. Since Brillouin readings could be recorded for strains as low as 10, very small volume change due to water absorption could be detected in a few minutes. Figure 9B shows the time rate of Brillouin signal changes during the hydrogel swelling and drying cycles of the 4-different water content clay hosts.

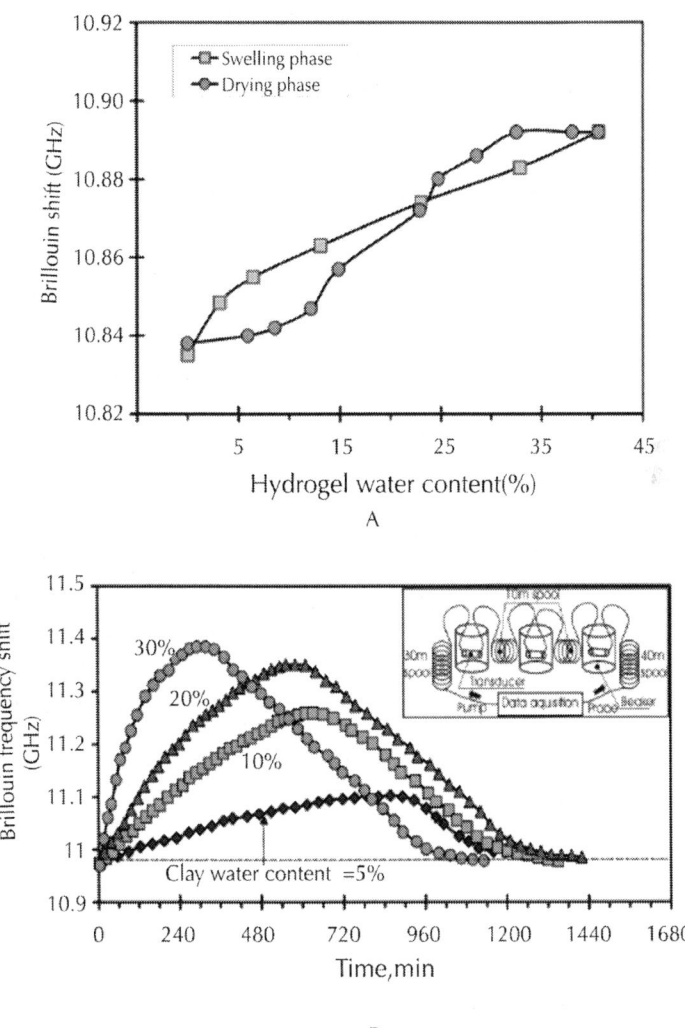

Figure 9: (A) Calibration of Brillouin Shift of fiber versus the water content by dry weight of the hydrogel disc. (B) Brillouin signal shift as a function of swelling/shrinkage time and clay water content (*Inset*– sketch of experimental set up).

In the experiments described above, the AEP60 polymer used would typically expand from 38 % to 400 % over dry volume when exposed to water. They are non-toxic and are manufactured in medical grades, approved for use in human wound care applications. This group of polymers does not swell in hydrocarbons and chlorinated solvents and has high thermal resistance. They are cross-linked to give them mechanical stability and accurate expansion characteristics. The water absorption and expansion factor can be accurately pre-defined at the formulation stage. Full expansion is reproducible over many wetting and drying cycles and is consistent over a wide range of pH and dissolved solid concentrations. A different integration of optical fiber and polymer transducer was used to improve the polymer response kinetics as shown in Figure 10. In this design, the polymer was reduced to smaller size discrete sleeves (1cm length x 0.3 cm diameter) that were bonded over the optical fiber. The bonding adhesive used was Locktite 414, a super bonding, quick drying adhesive containing cyanoacrylate and is intended for plastics and vinyl. The Locktite 414 was applied to each end of the polymer sleeve also.

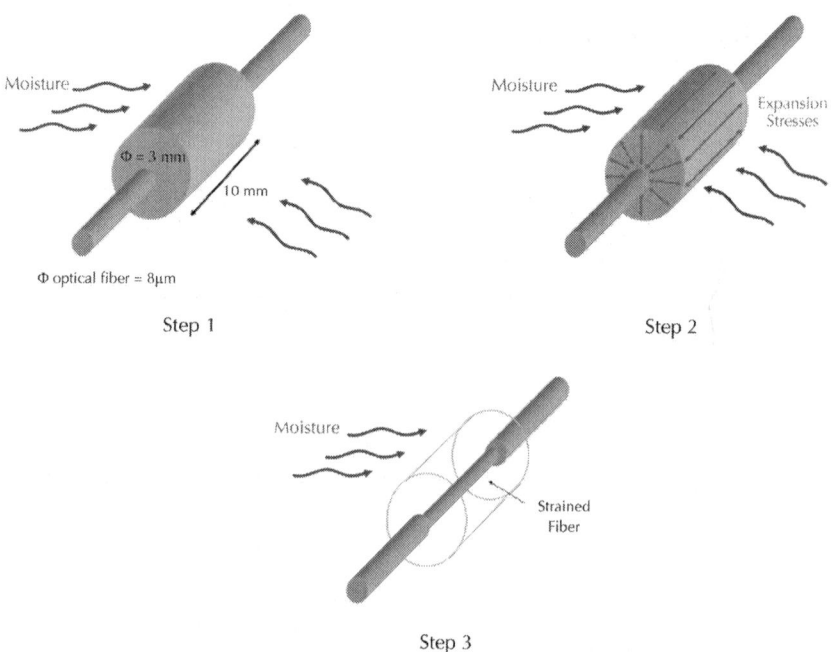

Figure 10: The working principle of the integrated sensor. Step 1 - Solution sorbed by the chemical selective polymer sleeve bonded onto the fiberoptic cable. Step 2 - The selective polymer swell upon encountering the target compound or ion in the solution. Step 3 - The swelling of the bonded polymer sleeve induces a "tangential pull" or "axial strain" on to the fiber locally.

The reduced size was anticipated to improve the swelling kinetics and alleviate the hysteresis affects observed in the previous configuration. The working principle of the integrated sensor is also depicted in Figure 10, where first the influx of the target substance (e.g. water) into the polymer transducer causes swelling of the bonded polymer. The swelling causes the bonding interface to strain and cause the fiber elongate in tangential pull. The fiber strain can then be recorded with location and amplitude, as shown in Figure 11, indicating where along the fiber line the influx of the target substance had occurred, and also the calibrated quantity of the substance based on the degree of swelling of the polymer, respectively.

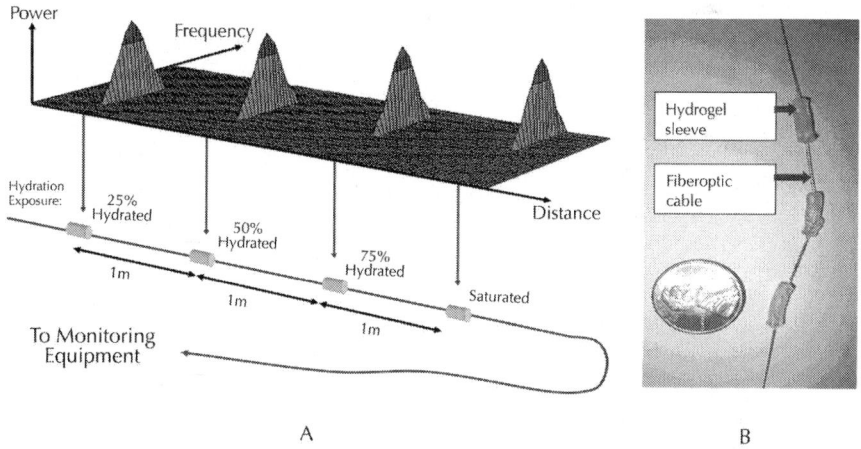

A   B

Figure 11: The operation schematic and the photograph of a spent water sensor: (A) Conceptual schematic of the integrated sensor assembly and its working principle as a distributed water sensor. (B) A laboratory prototype of the integrated water sensor.

Repeated swelling and shrinkage tests of the polymer sleeve component of the integrated water sensor showed hysteresis of length and diameter change. Both the length and diameter of the of the polymer sleeve expanded by 35 % (~ 0.35 cm and ~0.1 cm, respectively) after being soaked in water for three days. The majority of this expansion occurred during the first 12 hours of soaking. Increasing with every cycle, the final dry length of the polymer was greater than the original by ~ 0.025 cm (~2.5 %). As more cycles were completed, the diameter increased to its maximum faster, but the value of this maximum decreased. The final diameter of the polymer sleeve was fairly constant for each cycle, slightly less than the original by ~ 0.0005 cm (~0.2 %).

The magnitude and rate of swelling correlated directly with the initial water content when the polymer sleeves were embedded in test clay specimens of different water contents. Once again the full swelling occurred between 8 to 12 h when the sleeves were embedded in wet clay. Figure 12A shows that the polymer linear extension and clay water content relation was fairly linear. The linear relation is desirable for robust calibration. Figure 12B shows the dimensionless frequency shift response of the integrated water sensor to clay water content increase. The figure plots two spectrums, the shift spectrum at the location of the expanded polymer and a spectrum near the polymer location that does not undergo the swelling stress.

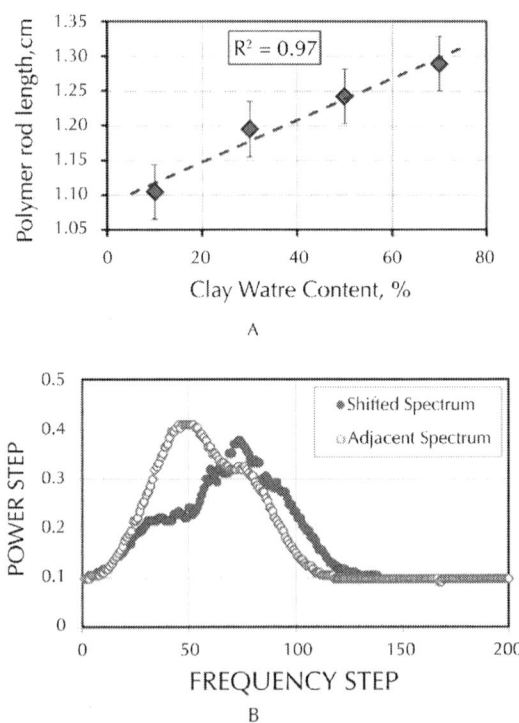

Figure 12: (A) Variation of embedded AEP60 polymer sleeve length vs clay w/c in swelling behavior. (B) Shifted frequency spectra for integrated sensor under swelling stresses with increased water.

The actual Brilloin frequency shift was measured 0.0432 GHz corresponding to axial strain of 0.098 % for the fiber. The actual elongation of the polymer rod was 3 mm, which fully transferred to the fiber would have corresponded to an average strain of 0.14 % for the fiber over the 12.6 ns pulse. This meant that about 70 percent of the axial tug generated by the swelling of polymer was transferred to the fiber. The other 30 percent can be attributed to slippage across the length of the bond between the fiber and the polymer sleeve or other process related inefficiencies.

These test results demonstrated the viability of integration of optical fiber with reactive polymer as a BOTDR based sensor. Yet, the slow kinetics of the response (~12 hr for full swelling in case of water) rendered the prototype assembly limited for quick detection and measurement purposes. Hence a new polymer and fiber-optic cable configuration is proposed where a thin layer (e.g. on the order of few hundred μm) of the reactive polymer is brushed and bonded onto the fiber-optic cable. Figure 13presents variations of the conceptual sensor where the reactive polymer coat is continuous. The working principle of this new configuration is similar to the previous ones described, all based on BOTDR, with the exception that the continuous coating of a thin layer reactive polymer is anticipated to provide a truly distributed and fast detection mechanism.

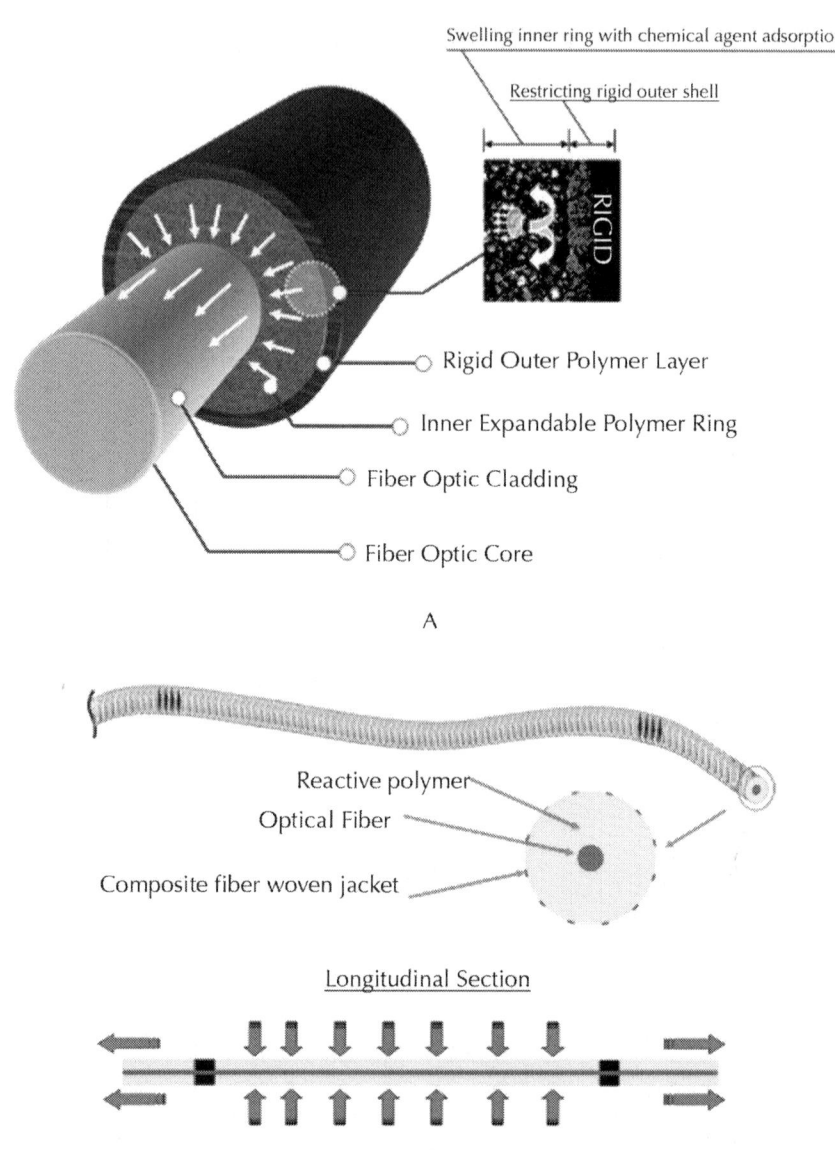

Figure 13: Advanced integrated chemical sensors based on BOTDR/BOT-DA: (A) Cross-section of integrated sensor with core/shell polymer layers; (B) Working principal of an integrated sensor with polymer/jacket combination.

Reactive Polymers Principles Used as Sensors

Point detection fiber optic sensors have been developed for measurement of liquid levels, chemical species (inorganic and organic), drugs, environmental agents (such as pollutants and pesticides), biochemical reactions, and to monitor a wide variety of various chemical processes (Wolfbeis, 2000). One of the major components of a sensor system is the sensing or recognition element. Polymers have often been utilized as a chemical sensing material. The interaction of an analyte with the polymer coating is of prime importance. A signal needs to be generated once this interaction occurs, and much work has been carried out to understand the interaction of the polymer coating with the analyte molecules, especially in terms of the diffusion behavior of the analyte through the polymer coating to the actual sensor itself and its subsequent enrichment within the polymeric coating layer (Philips, et al, 2003).

The integrated chemical sensor based on BOTDR discussed in here is a single SBS (Stimulated Brillouin Scattering) sensing optical fiber bonded with such a polymer, which swells selectively in the presence of a target chemical. The polymer coat is cross-linked to swell in a preferential direction. During swelling, the bonded polymer coat exerts a radial or tangential force at the polymer-fiber interface, hence an axial strain on the fiber. The SBS signals are generated along the fiber at the discrete points of chemical contact triggered by polymer swelling, hence the local axial strains in the fiber. Consequently, the location of the target chemical can be detected simply by linear positioning of the SBS signals over the entire length of the cable. Because the detection is based on local physical changes in the fiber and not on loss of transmitted light, widely distributed sensing is possible without high power requirements. The specific detection and measurement components of the integrated sensor described above include, a core/shell type (multi-layer) polymer coating, and an inner fiber optic cable core, or a rigid woven jacket conceptualized in Figure 13.

In this conceptual design, the outer cross-linked polymer coating (shell) serves as a rigid, high permeability filter and confinement to the inner polymer (core). The core is a flexible, chemically selective polymer, preferably with rapid mass sorption kinetics. While the highly networked rigid polymer shell confines and directs the cores welling toward the fiber optic cable, the selective detection of chemicals is based on the thermodynamics and kinetics of chemical sorption and swelling of the core polymer layer. A volume change or "swelling" occurs in the inner flexible polymer layer as a result of mass sorption of the surrounding analyte. Solute/solvent sorption interrupts the intermolecular forces between individual chains of a lightly cross-linked or a linear polymer adjacent to the glass fiber, resulting in swelling forces. The outer, permeable but rigid polymer layer comprised of highly networked cross-linked chains help confine the volume change tendency and direct part of the swelling forces inward (radially and/or longitudinally) thus creating a hoop or a tangential stress on the fiber at the polymer interface. This "pinch" or tug" of the fiber cause changes in the elastic and refractive properties of the fiber locally, generating a shift in its original Brillouin frequency at that local. Comparing the new frequency to the original, it is then possible to quantify the change in terms of the swelling forces and the quantity of the absorbed substance.

A multi-chemical sensor can be developed by bundling polymer-coated fibers of different functions and sensitivities. The entire sensor assembly can be several tens of kilometers of optical fiber hosting several 100 measurement locations on the same line, each at a minimum spatial resolution of 1-meter. The photonics assembly connected to one end of the fiber allows fast detection of discrete sensing locations rendering the entire assembly a multiplexed network of many point sensors on the same transmittal line. Hence, the fiber line coupled with the stringed transducers can be embedded linearly or laced into a host medium (i.e. water pipeline, paved surfaces, porous media such as soil or concrete) to detect target chemical(s) online over large distances, areas or spaces by linear positioning of the fiber.

Reactive Polymers

The molecularly imprinted polymers are often used to improve selectivity (Wolfbeis, 2000; Philips et al, 2003). The incorporation of desired functional monomers into the polymer structure further enhances the selectivity to a given analyte. These polymers are cross-linked and prepared by free radical polymerization processes such as solution or dispersion polymerization with acrylic or vinyl monomers. Polyelectrolyte gels are charged cross-linked three-dimensional networks of monomers that possess high swelling capability due to solvent sorption. The amount of swelling is known to be a string function of pressure, temperature, ion concentrations and pH changes (Siegel, 1993; Siegel et al, 1998; Matsuo and Tanaka, 1988). Their swelling and kinetics depend on parameters such as the degree of cross-linking (Skouri et al, 1995), external salt molarity (Yin et. al, 1992), and the degree of gel ionization rule (Katchalsky and Michaeli, 1995; Yin et al, 1992).

An emulsion or miniemulsion polymerization approach can be utilized to prepare film-forming polymer latexes with desired functional moieties to be used to coat optical fibers. The base latex polymer may be based on acrylic (e.g., n-butyl acrylate, n-butyl methacrylate) or styrene/acrylic film-forming compositions (i.e., with glass transition temperatures (T_g) of room temperature or lower). These latexes are prepared by conventional emulsion polymerization or by a miniemulsion polymerization process in the case where monomers with very low water solubility are used. In the miniemulsion polymerization process, the monomer would be emulsified in the presence of an aqueous surfactant (such as sodium lauryl sulfate) / costabilizer (e.g., hexadecane dissolved in the (co)monomer mixture) combination using a high shear device to form miniemulsion droplets which could then be polymerized in the presence of a free radical such as potassium persulfate.

The polymer would also be crosslinked to varying extents using crosslinking monomers such as ethylene glycol dimethacrylate, divinylbenzene, or bisacrylamide. In addition, functional monomers can be copolymerized along with the base acrylic or styrene/acrylic monomers. One type of monomer is a carboxylic acid such as methacrylic acid (MAA), which copolymerizes well with the base monomers. At high concentrations, this latex could function as an alkali-swellable latex whereby the latex particle size, and coating swellability, would increase dramatically upon neutralization in aqueous solutions of high pH (e.g. > 10) which would trigger a sensor response. N-methylol acrylamide (NMA) may also be incorporated into the base copolymer composition to obtain a crosslinked polymer, which can also act as a hydrogel, which could also swell when exposed to water. In addition, the monomer, N-(isopropylacrylamide) (NIPAM) will also be utilized for forming hydrogel particles which can swell when exposed to water. Incorporating NIPAM into a polymer composition would also lead to the formation of a thermosensitive polymer coating since poly(NIPAM) exhibits a strong phase transition above 32°C. It is also possible to copolymerize a alkoxysilane-containing monomer with the acrylic or styrene-acrylic monomers via miniemulsion polymerization to enhance the compatibility of the polymer coating with the glass optical fiber.

Bonding/Lamination

The extent of bonding of the polymer coating to the glass optical fibers is critical. The optical fibers can either be coated with the manufacturer's cladding removed or in place. Preliminary experiments have shown that it is difficult to coat the uncoated glass fibers. These fibers are brittle without the manufacturer's cladding in place; the composition of which is unknown. The fiber can be passed through a coagulant bath prior to its immersion in the latex bath. Similar to dip coating, which has been utilized in preliminary coating experiments, the latex will coagulate onto the glass fiber. The surface of the fiber needs to be made hydrophilic for this process. This can be achieved by either physical adsorption of nonionic water-soluble polymer such as poly(vinyl alcohol) (PVOH) or by corona treatment of the fiber surface. In addition, the polymer processing techniques used in wire coating applications can also be applied to the case of the optical fibers. Important coating parameters would include the solids content of the latex (a high solids content is needed to control the rheology of the dispersion to be coated; a reasonable viscosity is needed for effective coating). Latexes can also be made self-thickening by the incorporation of carboxyl groups into the latex particles. A thickener can also be added to a latex composition to adjust the coating viscosity.

In addition, the surface tension of the latex would need to be controlled to give good wetting onto the glass fiber. Contact angle measurements on glass substrates can be used to determine the optimum wetting behavior before moving on to the glass fiber itself. The thickness of the polymer coating would also need to be varied to determine the necessary thickness needed to give a good, measurable response when exposed to solvent or water containing the heavy metal ions. If the coating is not thick enough, the response to the analyte may be too weak. If the coating is not uniform on the fiber, there would be unexposed regions of the fiber which would affect the detection limit and sensitivity. In addition, there needs to be good adhesion of the coating to the fiber, otherwise delamination could occur. Silane adhesion promoters can be explored to enhance adhesion of the polymer coating onto the glass fiber substrate. The drying temperature and drying conditions (e.g., time and temperature that the latex-coated fibers are dried in an oven to ensure good film integrity or the use of forced heated air flow over the fibers) are also critical coating variables to be investigated.

Kinetics

The kinetics of swelling of the polymer coatings when exposed to aqueous or organic media needs to be evaluated by monitoring the changes in the dimensions of the coating or the gravimetric uptake of the media by the polymer. The time-dependent changes can be analyzed to give an idea of the best polymer architecture to obtain an optimum sensor response when exposed to a given chemical. The time constant for the sensor response needs to be determined and correlated with the swelling kinetics of the polymer coating to achieve the best sensor performance.

Development of a Prototype pH Sensor with Reactive Polymer Coating

The development of an optical fiber pH sensor based on hetero-core fiber structure coated with an acrylic polymer doped with Prussian blue is discussed here. In this design, the pH changes of the surrounding medium affects the Prussian blue present in the layer and produce a change in the refractive index of the layer. The pH changes are then observed as an increment in the hetero-core transmission signal.

Building of Hetero-Core Optical Fibers with Reactive Layer

The hetero-core fibers were constructed using two different length and two different types of optical fiber. In this case two types of single-mode fibers (SMA and SMB) and two of multimode fibers (MMA and MMB) were used. First, two pieces of MM fiber, stripped of its coating polymer (3 cm section) were spliced to a stripped SM fiber on each side. The hetero-core fibers were treated with Prussian Blue 0.1 mM

(PB), polyvinyl alcohol(PVOH) at 4 %,acrylic polymer emulsion (APE) at 50 % plus and their combinations, like PVOH + PB and APE + PB to develop a reactive coat over the stripped surfaces. A small U-shape container made of a glass capillary was fixed to a mechanical mount and was filled with the mixture of polymer support and Prussian blue sensitive material. Then the single-mode section of the hetero-core fiber was immersed for 5 minutes into the solution after which the fiber was removed and dried at room temperature. In this manner the sensitive material was adhered to the single-mode section of the hetero-core fiber.

One end of the hetero-core fiber was connected to a white light source Yokogawa AQ4305 and the other to the spectrum analyzer Ando AQ6315A (Figure 14A). The set-up was used to measure the transmission light during the modification process of the fiber and later to measure the response of the modified fiber to pH changes. In order to test the sensitivity of the device to changes in pH, a test was designed which consisted of immersing the optical fiber section modified with PVOH/PB or APE/PB in a Petri dish where the pH was varied by adding 0.1 M NaOH or 0.1 M HCl, recording each transmission spectrum changes in the wavelength range from 350 nm to 1700 nm (Figure 14B).

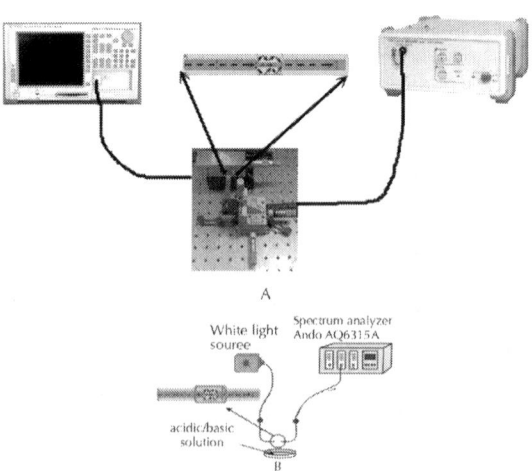

A

White light source

Spectrum analyzer Ando AQ6315A

acidic/basic solution

B

Figure 14: Set up to test prototype hetero-core fibers for chemical detection: (A) Diagram of the set-up to measure the transmission signal of the hetero-core fibers; (B) Diagram of the test system to determine the sensitivity of hetero-core fiber to pH changes.

The transmission spectra of two hetero-core fibers with PVOH/A (5 mm and 10 mm sections) were measured in different pH solutions are shown in Figure 14A and 14B, respectively. As seen in figure 15 the device has good sensitivity (-1.5 dB and -2 dB approximately), however the signal is erratic and not repeatable for different pH changes. This was attributed to solubility of PVOH in acidic conditions, and checked visually and with the transmission spectrum analysis. A new polymer, acrylic polymers emulsion (APE) was selected to replace PVOH. This polymer has similar characteristics as PVOH. It is water soluble, inexpensive, and colorless when dried, and has been reported as a good support in manufacturing of modified electrodes for pH determination.

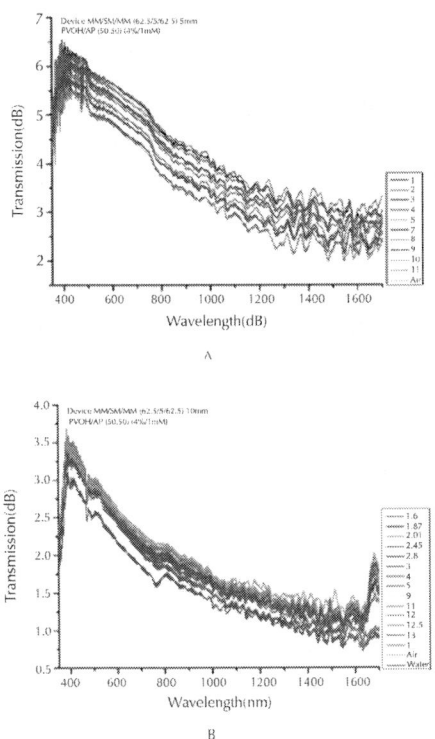

Figure 15: Transmission spectra of hetero-core fiber with PVOH/PB and its sensitivity to pH change with 5 mm (A) and 10 mm (B) length.

The transmission spectra of two hetero-core fibers with APE/A (5 mm and 10 mm sections) were measured in different pH solutions are shown in Figure 16A and 16B, respectively. As seen in Figure 16A, the device showed very obvious changes when subjected to acidic and basic conditions. The presence of three peaks in visible region, 400, 500 and 700 nm wavelengths were noted when optical fiber was in air. These signals were attributed to light absorption and loss of light by index refraction changes by the composite material. In acidic pH values the light losses were in the range of 1.5 to 3 dB while for basic pH levels transmission near to 0.5 dB.

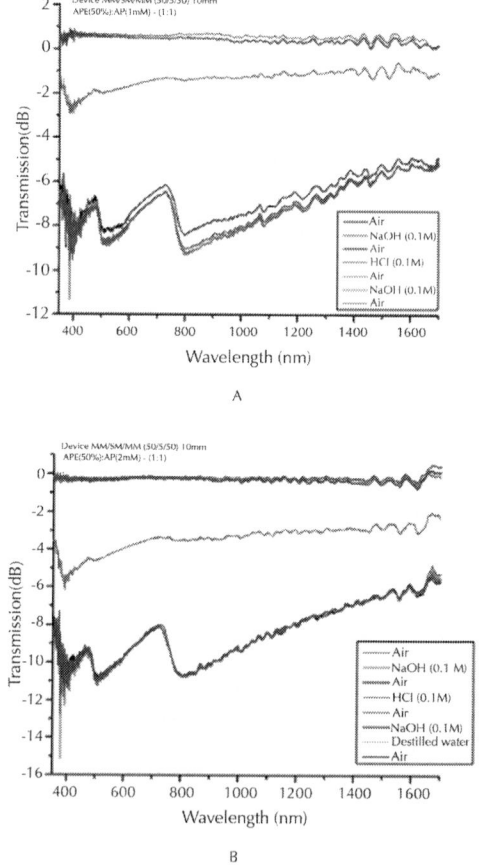

Figure 16: Transmission spectra of hetero-core fiber with APE/PB and its sensitivity to pH change with 1 mM (A) and 2 mM (B) PB.

In order to identify the origin of signals found, the concentration of Prussian blue was increased to 2 mM. As we can see in Figure 16B, the intensity of transmission peaks at 400, 500 and 700 nm was increased, suggesting that they are due to increased concentration of PB in the composite. It also shows that the device sensitivity increased from 1.5 to 3 dB with 1mM concentration of PB until 4 to 6 dB with 2 mM concentration of PB for acidic pH solutions, but behaved same as previous in basic solutions. Finally there was a good return to initial conditions after each change of interface (Figure 17).

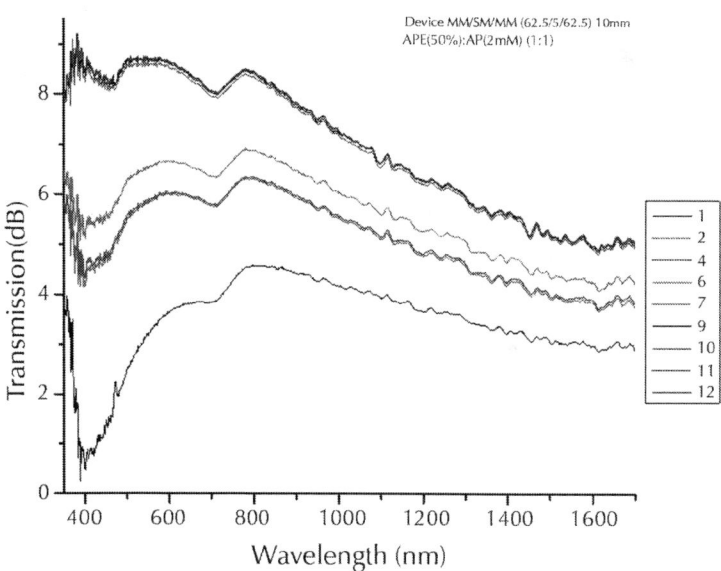

Figure 17: Transmission spectrum normalized to air of the modified hetero-core device with APE/PB (2 mM) and its sensibility to pH changes.

Subsequently we performed a sensitivity analysis for pH changes by taking the APE/PB modified fiber signal in air to use as a normalizing reference. As shown in Figure 18, there is good sensitivity to pH values lower than 7 with gains up to 6 dB at 400 nm (absorption or loss peak), whereas above pH 7 the peak is inverted, turning in a gains peak which may be due to the hydration process of the polymer and breaking of complex of PB by hydration. To identify the changes in transmission spectrums, the most characteristic signals (400, 700 and 800 nm) and the response to 1500 nm (common wavelength in telecommunication systems) were plotted independently.

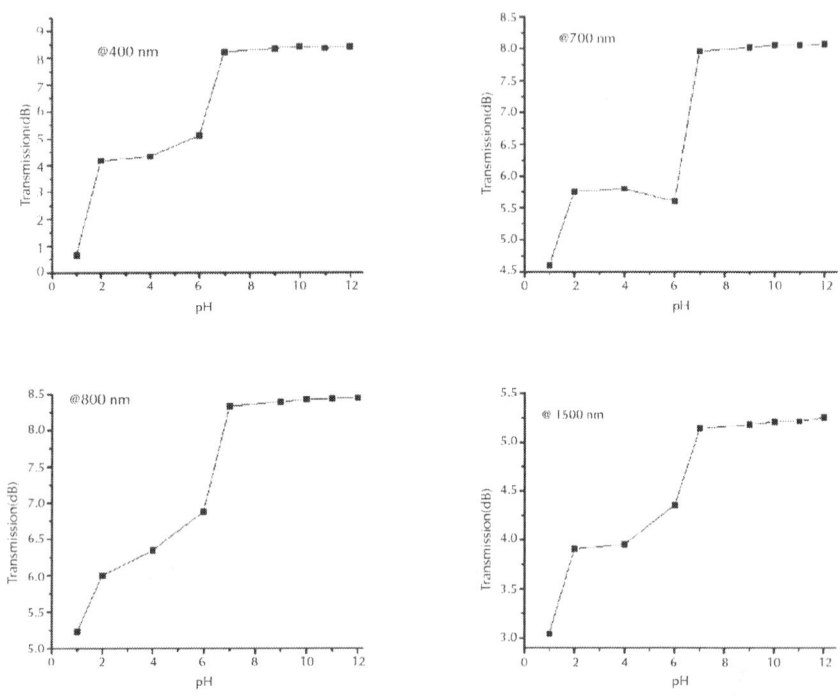

Figure 18: Transmission spectrum of hetero-core device modified with APE/PB and its sensibility to pH changes at 400, 700, 800 and 1500 nm wavelength.

Analyzing the charts of Figure 18 shows that the transmission intervals are decreasing with increasing wavelength, which demonstrates good sensitivity of the device. As previously mentioned, the pH changes were more evident at pH values less than 7, due to that pH values higher than 7, it promotes the process of hydration of Prussian blue complex (Equation 4) and the signal grows weak on each pH change (García – Jareño et al, 1996).

$$Fe_4\left[Fe(CN)_6\right]_3 + 12OH^- \Leftrightarrow 4Fe(OH)_3 + 3\left[Fe(CN)_6\right]^{4-} \qquad (4)$$

Based on the results obtained so far, the 10 mm rather than the 5 mm length hetero-core device is recommended since their sensitivity and the evanescent wave field is bigger than 5 mm length devices. Also the 10 mm length device provides a gain of about 2 dB at throughout the analysis spectrum.

Actually, all devices showed in this paper are preparing to their application in real scenarios, with the intention to quantify physicochemical properties directly to polluted soil without extraction from the field and pre-treatment of sample, which could reduce time and costs of analytical determination, increasing the sensibility, detection and quantification limits in comparison with spectroscopic and spectrometric techniques, to take the best professional decision to remediate in the better technical conditions the polluted soil.

CONCLUSIONS

For pollution detection and soil remediation purposes it is essential to have relevant and reliable information on the soil structure, the hydrogeological circumstances and accumulation zones of the detected pollutants. Combined application of geological, hydrogeological and geophysical investigations prior the placement of the optical fiber in the field may increase the efficiency of the monitoring technique.

Spatially resolved mapping of chemical constituents is an important need in a variety of environmental and geo-environmental applications. For example, spatially resolved analyte monitoring can simultaneously indicate and locate when an accepted level of exposure to toxic or explosive species has been exceeded, and can track its source.

The capability of long-range distributed sensing is unique to optical – fiber technology. A distributed fiber optic sensor returns a value of a target measurement as a function of the linear position along the fiber length. The only contact between the point to be measured and the observation area is the optical fiber.

ACKNOWLEDGEMENTS

The authors would like to thank the Consejo Nacional de Ciencia y Tecnología de los Estados Unidos Mexicanos (CONACyT), L'Oreal,

Academia Mexicana de Ciencias (AMC) and Fundación México – Estados Unidos para la Ciencia (FUMEC).J. A. García is grateful to CONACyT for his scholarship.

REFERENCES

1.	Abdi, M. M., Abdullah, L. Ch., Sadrolhosseini, A. M., Yunus, W. M. M., Moksin, M. M., Tahir, P. M. (2011). Surface plasmon resonance sensing detection of mercury and lead ions based on conducting polymer composite, Plos One, 6, e24578.

2.	Achatz, D. E., Ali, R., Wolfbeis, O. S. (2011). Luminescent chemical sensing, biosens- ing, and screening using upconverting nanoparticles, Top Current Chemistry, 300, 29 – 50.

3.	Aksuner, N. (2011). Development of a new fluorescent sensor based on a triazolothiadiazin, derivative immobilized in polyvinyl chloride membrane for sensitive de- tection of lead (II) ions, Sensors and Actuators B, 157, 162 – 168.

4.	Anastasio, S., Pamukcu, S., Pervizpour, M. (2007). BOTDR Detection of Chemical & Liquid Content, Proc. of the 7th FMGM, GSP 175 Int. Sym. on Field Measurements in Geomechanics, ASCE, Boston, MA, 1 - 12.

5.	Anastasio. S, Pamukcu, S., Pervizpour, M. (2007). Chemical Selective BOTDR Sensing for Corrosion Detection on Structural Systems, Chang, ed., Proc. of the 7th Int. Work- shop on Structural Health Monitoring (IWSHM 2007), Stanford, CA, 1701 - 1708.

6.	Antico, E., Lerchi, M., Rusterholz, B., Achermann, N., Badertscher, M., Valiente, M., Pretsch, E. (1999). Monitoring Pb2+ with optical sensing films, AnalyticalChimicaActa, 388, 327 - 338.

7.	Balaji, T., Sasidharan, M., Matsunaga, H. (2006).Naked eye detection of cadmium us- ing inorganic–organic hybrid mesoporous material, Analytical Bioanalytical Chemis- try, 384, 488-494.

8.	Bao, X., DeMerchant, M., Brown, A., and Bremner, T. (2001). Tensile and compressive strain measurement in the lab and field with the distributed Brillouin scattering sen- sor.Journal of Lightwave Technology, 19, 1698.

9. Bao, X., Dhliwayo, J., Heron, N., Webb, D. J., Jackson, D. A. (1995). Experimental and theoretical studies on a distributed temperature sensor based on Brillouin scattering, Journal of Light Technology, 13, 1340 - 1348.

10. Buerck, J., Roth, S., Kraemer, K., Mathieu, H. (2001). OTDR distributed sensing of liq- uid hydrocarbons using polymer-clad optical fibers, Proceedings of The Second In- ternational Symposium and Workshop on Time Domain Reflectometry for Innovative Geotechnical Applications, C. H. Dowding, ed. Academic, Evanston, Il., 496 - 509.

11. Cui, Q., Pamukcu, S., Xiao, W., Guintrand, C., Toulouse, J., Pervizpour, M. (2009). Distributed fiber sensor based on modulated pulse base reflection and Brillouin gain spectrum analysis, Applied Optics, 48 (30), 5823 - 5828.

12. Cui, Q., Pamukcu, S., Lin, A., Xiao, W., Toulouse, J. (2010). Performance of double side band modulated probe wave in BOTDA distributed fiber sensor, Microwave and Optical Technology Letters, 52, 2713 - 2717.

13. Cui, Q., Pamukcu, S., Lin, A., Xiao, W., Herr, D., Toulouse, J., Pervizpour, M. (2011). Distributed temperature sensing system based on Rayleigh scattering BOTDA, IEEE Sensors Journal, 11(2), 399 - 403.

14. Cui, Q., Pamukcu, S., Xiao, W., Pervizpour, M. (2011). Truly distributed fiber vibra- tion sensor using pulse base BOTDA with wide dynamic range, IEEE Photonics Tech- nology Letters, 3 (24), 1887 - 1889.

15. Fellay, A., Thévenaz, L., Facchini, M., Niklès, M., Robert, P. (1997). Distributed sens- ing using stimulated Brillouin scattering: towards ultimate resolution, Optical Fiber Sensors, OSA Technical Digest Series, Optical Society of America, Washington, D.C. 16, 324 – 327.

16. Fen, Y. W., Mahmood, W., Yunus, M., Yusof, N. A. (2012). Surface plasmon reso- nance optical sensor for detection of Pb2+ based on immobilized p-tert-butylcalix[4]arene-tetrakis in chitosan thin film as an active layer, Sensors and Actuators B, 171 – 172, 287 – 293.

17. Fen, Y. W., Yunus, W. M. M. (2013). Utilization of chitosan - based sensor thin films for the detection of lead ion by surface plasmon resonance optical sensor, IEEE Sen- sors Journal, 13, 1413 - 1418.

18. Fen, Y. W., Yunus, W. M. M., Talib, Z. A. (2013). Analysis of Pb(II) ion sensing by crosslinked chitosan thin film using surface plasmon resonance spectroscopy, Optik, 124, 126 – 133.

19. Forzani, E. S., Foley, K., Westerhoff, P.,Tao, N. (2007). Detection of arsenic in ground- water using a surface plasmon resonance sensor, Sensors and Actuators B, 123, 82 – 88.

20. García-Jareño, J. J., Navarro-Laboulais, J., Vicente, F. (1996). Electrochemical Study of Nafion Membranes / Prussian Blue Films on ITO Electrodes. Electrochimica Acta, 41, 17, 2675 – 2682.

21. Galindez-Jamioy, C. A., López-Higuera, J. M. (2012). Brillouin Distributed Fiber Sen- sors: An Overview and Applications. Journal of Sensors, 204121, pp 17.

22. Grattan, K. T. V., Meggitt, B. T. (1999). Optical Fiber Sensor Technology, Chemical and Environmental Sensing, KlumerAcademinc Publishers, Vol. 4.

23. Guillemain, H., Rajarajan, M., Sun, T., Grattan, K. T. V. (2009). A self-referenced re- flectance sensor for the detection of lead and other heavy metal ions using optical fi- bres, Measurement Science Technology, 20, 045207.

24. Guo, L., Zhang, W., Xie, Z., Lin, X., Chen, G. (2006). An organically modified sol–gel membrane for detection of mercury ions by using 5,10,15,20-tetraphenylporphyrin as a fluorescence indicador, Sensors & Actuators B, 119, 209 - 214.

25. Horiguchi, T., Shimizu, K., Kurashima, T., Tateda, M., Koyamada, Y. (1995). Devel- opment of a distributed sensing technique using Brillouin scattering, Journal of Light- wave Technology, 13, 1296 – 1302.

26. Jerónimo, P., Araújo, A., Conceição, B. S. M., Montenegro, M. (2007). Optical sensors and biosensors based on sol-gel films, Talanta, 72, 13 – 27.

27. Kasik, I., Mrazek, J., Martan, T., Pospisilova, M., Podrazky, O., Matejec, V., Hoyero- va, K., Kaminek, M. (2010) Fiber-optic pH detection in small volumes of biosamples, Analytical and Bioanalytical Chemistry, 398, 1883 – 1889.

28. Katchalsky, A., Michaeli, I. (1995). Polyelectrolyte gels in salt solutions, Journal of Pol- ymer Science, 15 (69).

29. Kee, H. H., Lees, G. P., Newson, T. P. (2000). All-fiber system for simultaneous inter- rogation of distributed strain and temperature sensing by spontaneous Brillouin scat- tering.Optics Letters, 25, 695.

30. Kocincova, A., Borisov, S., Krause, C., Wolfbeis, O. (2007).Fiber-optic microsensors for simultaneous sensing of oxygen and pH, and of oxygen and temperature.Analyti- cal chemistry, 79, 8486 – 8493.

31. Krohn, D. A. (1988).Fiber Optic sensors: Fundamental and applications, Instrument Society of America.

32. Lee, S. -H., Kumar, J., Tripathy, S. K. (2000). Thin film optical sensors employing pol- yelectrolyte assembly. Langmuir, 16.

33. Lin, T. –J., Chung, M. –F. (2009). Detection of cadmium by a fiber-optic biosensor based on localized surface plasmon resonance, Biosensors and Bioelectronics, 24, 1213 – 1218.

34. Maier, S. A. (2007). Plasmonics Fundamental and applications, Springer.

35. Matsuo, E. S., Tanaka, T. (1988). Kinetics of discontinuous volume phase transition of gels, Journal of Chemical Physics, 89 (3): 1695.

36. Mayra, T., Klimant, I., Wolfbeis, O. S., Werner, T. (2008). Dual lifetime referenced op- tical sensor membrane for the determination of copper (II) ions, Analytical ChimicaAc- ta, 462, 1 - 10.

37. McDonagh, C., Burke, C. S., MacCraith, B. D. (2008). Optical Chemical Sensors, Chem- ical Review, 108, 400 - 422.

38. Ohno, H., Naruse, H., Kihara, M., Shimada, A. (2001) Industrial applications of the BOTDR optical fiber strain sensor, Optical Fiber Technology, 7,45 – 64.

39. Orellana, G., Haigh, D. (2008) New trends in fiber-optic chemical and biological sen- sors. Current Analytical Chemistry, 4, 273 – 295.

40. Pamukcu, S., Cetisli, F., Texier, S., Naito, C., Toulouse, J. (2006). Dynamic strains with Brillouin scattering distributed fiber optic sensor, GeoCongress 2006, 187, ASCE pp. 31 - 36.

41. Pamukcu, S., Texier, S., Toulouse, J. (2006).Advances in water content measurement with distributed fiber optic sensor, GeoCongress 2006a, 187, ASCE, pp. 7 - 12.

42. Phillips, C., Jakusch, M., Steiner, H., Mizaikoff, B., Fedorov, A. G. (2003).Model-based optimal design of polymer-coated chemical sensors, Analytical Chemistry, 75, 1106 - 1115.

43. Prabhakaran, D., Nanjo, H., Matsunaga, H. (2007). Naked eye sensor on polyvinyl chloride platform of chromo-ionophore molecular assemblies: A smart way for the colorimetric sensing of toxic metal ions, Analytical ChimicaActa, 601, 108 - 117.

44. Siegel, R. A. (1993). Hydrophobic weak polyelectrolyte gels: studies of swelling equi- libria and kinteics, Advanced Polymer Science, 109, 233.

45. Siegel, R. A., Falamarzian, M., Firestone, B. A., Moxley, B. C. (1988).pH-controlled re- lease from hydrophobic/polyelectrolyte copolymer hydrogels, Journal of Controlled Release, 8, 179.

46. Skouri, R., Schosseler, F., Munch, J. P., Candau, S. J. (1995). Swelling and elastic prop- erties of polyelectrolyte gels, Macromolecules, 28, 197.

47. Texier, S., Pamukcu, S., Toulouse, J. (2005). Advances in subsurface water-content measurement with a distributed Brillouin scattering fibre-optic sensor, Proc.of SPIE 5855, 17th Int. Confer. On Optical Fibre Sensors, OFS-17, Bruggs, Belgium, pp. 555 - 558.

48. Texier, S., Pamukcu, S., Toulouse, J., Ricles, J. (2005).Brillouin scattering fiber optic strain sensor for distributed applications in civil infrastructure, Chang, ed., 5th Int. Workshop on Structural Health Monitoring (IWSHM 2007), Stanford, CA, pp. 1395 - 1402.

49. Turel, M., Pamukcu, S. (2006) Brillouin scattering fiber optic sensor for distributed measurement of liquid content and geosynthetic strains in subsurface, Geoshanghai GSP: Site and Geomaterial Characterization, ASCE, Shanghai, PRC, pp. 72 - 79.

50. Villatoro, J., Monzón-Hernández, D. (2006). Low-cost optical fiber refractive-index sensor based on core diameter mismatch. Journal of Light-wave Technology, 24, 1409 - 1413.

51. Wolfbeis, O. (2008). Fiber-optic chemical sensors and biosensors. Analytical chemistry 80, 4269 – 4283.

52. Wolfbeis, O. S. (2000). Fiber-Optic Chemical Sensors and Biosensors, Analytical Chem- istry, 72, 81R.

53. Yin, Y. L., Prud'homme, R. K., Stanley, F. (1992). Chapter 6:

Relationship between poly(acrylic acid) gel structure and synthesis. In A.J Harland and R.K. Prud'homme, editors, Polyelectrolyte Gels. ACS Symp. Series 480, ACS Washington D.C.

54. Yusof, N. A., Ahmad, M. (2003). A flow-through optical fibre reflectance sensor for the detection of lead ion based on immobilized gallocynine, Sensors and Actuators B, 94, 201 – 209.

Citations

CHAPTER 1

Nosheen Mirza, Qaisar Mahmood, Mohammad Maroof Shah, Arshid Pervez, and Sikander Sultan, "Plants as Useful Vectors to Reduce Environmental Toxic Arsenic Content," The Scientific World Journal, vol. 2014, Article ID 921581, 11 pages, 2014. doi:10.1155/2014/921581.

CHAPTER 2

S Luke Flory, Kimberly A Lorentz, Doria R Gordon, and Lynn E Sollenberger, Experimental Approaches for Evaluating the Invasion Risk of Biofuel Crops, doi:10.1088/1748-9326/7/4/045904.

CHAPTER 3

Terrence Thomas and Cihat Gunden, Organic Agriculture, Sustainability and Consumer Preferences, doi: 10.5772/58428.

CHAPTER 4

Beata Feledyn-Szewczyk, Jan Ku , Krzysztof Jo czyk, and Jarosław Stalenga (2014), The Suitability of Different Winter and Spring Wheat Varieties for Cultivation in Organic Farming, Organic Agriculture Towards Sustainability, Prof. Vytautas Pilipavicius (Ed.), ISBN: 978-953-51-1340-9, InTech, DOI: 10.5772/58351.

CHAPTER 5

Farzin Shabani, Lalit Kumar, and Atefeh Esmaeili, Use of CLIMEX, Land use and Topography to Refine Areas Suitable for Date Palm Cultivation in Spain under Climate Change Scenarios, doi: 10.4172/2157-7617.1000145.

CHAPTER 6

Sijuwade Adebayo and Idowu O Oladele (2014)., Organic agricultural practices among small holder farmers in South Western Nigeria, Organic Agriculture Towards Sustainability, Prof. Vytautas Pilipavicius (Ed.), ISBN: 978-953-51-1340-9, InTech, DOI: 10.5772/57598.

CHAPTER 7

Juan C. Durán–Álvarez and Blanca Jiménez–Cisneros (2014). Beneficial and Negative Impacts on Soil by the Reuse of Treated/ Untreated Municipal Wastewater for Agricultural Irrigation – A Review of the Current Knowledge and Future Perspectives, Environmental Risk Assessment of Soil Contamination, Dr. Maria C. Hernandez Soriano (Ed.), ISBN: 978-953-51-1235-8, InTech, DOI: 10.5772/57226.

CHAPTER 8

I. Robles, J. Lakatos, P. Scharek, Z. Planck, G. Hernández, S. Solís, and E. Bustos (2014). Characterization and Remediation of Soils and Sediments Polluted with Mercury: Occurrence, Transformations, Environmental Considerations and San Joaquin's Sierra Gorda Case, Environmental Risk Assessment of Soil Contamination, Dr. Maria C. Hernandez Soriano (Ed.), ISBN: 978-953-51-1235-8, InTech, DOI: 10.5772/57284.

CHAPTER 9

J. A. García, D. Monzón, A. Martínez, S. Pamukcu, R. García, and E. Bustos, Optical Fibers to Detect Heavy Metals in Environment: Generalities and Case Studies, http://dx.doi.org/10.5772/57285.

Index